Relativity Revealed
A Concrete Approach You Can Understand

Ray Christensen Jones

Emeritus Professor of Physics
Southwestern Oklahoma State University
Weatherford, OK 73096

Relativity Revealed
A Concrete Approach You Can Understand

Jones, Ray C., Ph.D. 1941 - 2007

Printed in the United States of America.

DEDICATION

This book is dedicated to the memory of its author, Ray Christensen Jones, Ph.D., our friend, colleague and peerless teacher who believed that anyone can understand this subject. We believe that this book, which was his last major work, will prove his point. Most of us were privileged to have heard the public lectures that evolved into this book. Printed words in black and white cannot convey the inimitable style and humor of Prof. Jones at his best. Nevertheless, they are his words and we believe them to be well worth the time spent in reading and pondering them. We believe that all who read them will agree with us. We hope, as he did, that this includes many young people who read this clearly written explanation of one of the fascinating features of the physical universe.

Jill Jones, Dean Chapman, Terry Goforth, Benny Hill, Stan Robertson

Acknowledgements

This book exists because those who admired, loved, and respected Ray Jones joined together to make it possible. Stan Robertson helped to locate the final versions of the chapter files and then spent many hours getting them into the correct format, with illustrations and equations all in their proper places. Dr. Benny Hill offered publishing advice and contacts with former students and helped to establish a connection with Tonya Shook, whose invaluable assistance led to the book's physical existence. Generous offers to proofread the text came from Dr. Wayne Trail and one of Ray's favorite students, Dr. L. Dean Chapman. Special thanks to Kinley Jones for the help she gave her father in constructing the spaceship figures. Many thanks to all who worked to make *Relativity Revealed* a reality.

Jill T. Jones

Table of Contents

Mathematics:

Preface

Everyone has heard of relativity, and secretly or not, many would like to know something about it, if for no other reason than to simply feel a little superior to the others in their class or car pool. However, the word "relativity" is almost synonymous with "extremely difficult to understand" or "weird," so fear of failure can prevent some from even trying to learn what it is about.

This book is intended for the general reader who has some curiosity about the universe in which we live. It is especially well-suited for ninth graders and up who might become engineers, physical scientists, or mathematicians. I know that I would have been thrilled to have had such a book when I was that age. The appendices can help improve math skills, and the development in the body portion will exercise logic abilities.

Now retired, I was a physics professor at a regional university for 32 years. During that time I taught almost all of the physics courses that physics majors take. I always tried to explain physics concepts verbally, not just relying on the mathematics. After the students had graduated and moved on to gainful employment, a number of them expressed their appreciation for this type of presentation. I greatly enjoyed teaching all levels of physics and engineering courses, but another love of my life is astronomy. I taught a freshman-level introductory astronomy class about sixty times. Relativity is only a small part of this course, but in dealing with these general students I improved my skills at presenting scientific concepts without using highly abstract methods.

I also love relativity and wanted to present it to general audiences, so in 1984 and again in 2003, I presented a series of free lectures on relativity for the general public. Attendance was very good, and attendees' comments were very encouraging. With these lectures, I honed my presentation of this material. This book has grown from the notes that were handed out at these seminars, not from my physics lecture notes. I have greatly enjoyed working up the presentation methods used here. I think that it shows in the book.

There are already enough books that deal with Einstein's life and how he might have arrived at the ideas of relativity. This book will deal with the subject of relativity only. Very few personalities will be mentioned, and then only for good reasons.

You might say to yourself, "Sure, I've heard of relativity all my life, but I have no idea what it is. Can I really learn something about it and actually be able to discuss it?" If you have picked up this book and read this far, then the answer is probably, "Yes." However, relativity cannot be learned by simply jumping into the middle; an organized approach to the subject must be followed. The limited exposure to relativity that people get from television, magazines, and the internet seldom provides such an organized approach. **Relativity Revealed** starts from a simple beginning, builds, and shows that much of relativity *is* accessible to the reading public.

Rather than only describing the subject, **Relativity Revealed** follows a guided discovery approach and actually derives relativity from the simple postulates used by Einstein. (A postulate is an assumed truth for the sake of consideration or argument, to see what logically follows from it.) Most of what we will develop comes from two elementary ideas. The first is very familiar: You have probably noticed that in a closed airplane or car nothing within seems any different because of your motion (as long as the ride is smooth, straight and at constant speed); flying feels the same as being parked on the ground. Postulate 1 is simply that this remains true no matter what the speed. The second postulate follows easily from astronomical observations of binary stars. These are very common star systems that consist of two stars in orbit around each other. Even though these systems might be so distant from Earth that their light must travel for hundreds of years to reach us, they appear to be following the same laws of motion as similar systems that are quite close to us. This tells us something wonderfully simple and important about light, that is, light's speed is not affected by the motion of its source. This is our second postulate. Later, when we look into gravity, we will add a third postulate which is simply the g-forces idea that acceleration and gravity feel the same. Putting these simple ideas together produces the theory of relativity.

Most of this book deals with what is called "Special Relativity." This follows from the first two postulates and includes results such as the slowing of time, the contraction of lengths, the conclusion that nothing can go faster than light, the most famous equation in the world, and other less familiar topics. We then add our third postulate that acceleration (change in velocity) and gravity feel the same, and then we can develop quite interesting conclusions about gravitation. For example, clocks, and time itself, run faster on the upper floors of a building than on the ground floor. We will briefly describe two fundamental gravity theories, including Einstein's general theory and its prediction of black holes. Another more recent theory of gravity is briefly described. Additional topics of interest that supplement relativity are also included: the Sun's energy source and energy in general, Doppler radar (weather and police), prospects for interstellar and time travel, colors, the visual appearance of rapidly moving objects, current relativity experiments, and relativity humor in the form of small jokes and limericks. There are some discussions of quantum (microscopic) considerations that affect such topics as time travel. Included are even a few questions and puzzles to encourage you to think a little more about the topic at hand. Answers and explanations are in the appendices.

Nature is mathematical; this fact is perhaps the most important discovery of the ancient Greeks. Relativity is a part of nature, and it cannot be adequately described or developed without the use of some simple equations to keep up with the logic that is involved. To quote Einstein, "Everything should be as simple as possible – but not simpler." We will discover relativity using gedanken experiments (German for thought experiments). These will be drawings of situations to which we will apply our postulates and/or earlier results in order to arrive at some new truth. Simple algebra makes this discovery process possible. The primary logical arguments that we will be following are conclusions from the drawings such as "This length plus that length equal a third length." Some of these lengths will be simple lengths, while others will be a speed (rate) times a time or even be a contracted length (a relativistic effect that we will discover early on). These steps are all printed in the book; you do not have to come up with anything on your own.

When we try to apply what we know about relativity to a new situation, there may be several already-discovered effects to apply. It may be that effect 1 and effect 2 will increase some new result, while effect 3 will decrease it. We cannot conclude from this alone that the net effect will be to increase the new result, because effect 3 might be the strongest of the three. The only way to keep up with the logic is to use the adding and multiplying that we already know. The early portion of the appendix will provide you with all that you need to know about math to successfully follow any of the logic in this book. Increasing your ability to use math might be the biggest practical payoff of all for reading this book.

In trying to arrive at some new result in relativity, it can happen that we must take some earlier result, say some time interval, and insert this into another equation. Students sometimes call this process "plug and chug." The chugging (simplification) has nothing to do with relativity directly, but always seems to result in a surprisingly simple conclusion. These simplifications are shown in the appendices so as not to clutter the development of the subject of relativity as presented in the body of the book. These manipulations are explained in great detail, step-by-step. Some may appear intimidating at first. Don't look past the first few pages of the appendix now! You might become discouraged. If you wait until it is suggested that you look at a certain section, then it will not seem so intimidating. These simplifications are presented only for those who want to see them. Your knowledge of relativity will not be limited by not studying them. Another reason for showing these simplifications is to completely derive all that is covered in this book on the subject of special relativity. We will not try for this lofty goal in all of the gravity discussions in Chapter 5. The math level would be too involved for the intended audience of this book.

This book can be read on several levels. One way is to read only the chapters and look at the diagrams and the simple equations from the diagrams in the body portion. Another method is to go through some of the math manipulations that are done in the appendices. A third method is, in addition, to read the more advanced optional topics that are treated in the appendices. Regardless of how the book is read, it cannot be read to good effect at the speed you would read a novel. One will have to stop and reflect for a moment or two, or even re-examine a drawing each time something is not immediately grasped; this is normally true for technical subject matter. *Relativity Revealed* is written with the aim of keeping technical language to a minimum. This book is essentially self-contained in that any technical terms or concepts that must be used are defined and described. Some of the most famous results from relativity come from physics. Topics such as energy, momentum, and mass will be described since they are necessary to understand these most famous results. Fortunately, everyone already has a good start on understanding these concepts. There will be essentially no attempt to bring in any physics content that is not necessary or desirable for the understanding of our principal topic, relativity.

Some relativity books intended for the general public seem to glorify the non-intuitiveness (weirdness) of certain aspects of relativity. *Relativity Revealed* takes a "nuts and bolts" approach; it does not become philosophical. It tries to make the subject concrete, rather than abstract. Relativity is inherently interesting and fun. It deals with the concepts of space and time and the consequences of these on various phenomena.

Fascinating conclusions are discovered frequently, which should keep the pages turning. Some of the conclusions will seem non-intuitive at first, but they will gradually become more natural to you as your knowledge and understanding of relativity grows. May your voyage of discovery into the wonders of relativity be mind-expanding and pleasant.

RCJ 11/25/06

Chapter 1 - The First Steps in Discovering Relativity

The huge Questorian spaceship first arrived on Earth in 1871. The Questor's primary mission had been to peacefully explore the galaxy, but their government financing for this project had been curtailed. To provide the needed financial resources, they began offering a transportation service to the pleasure planet, Blithe, which was about one hundred light years from Earth. Blithe had all of the attractions that anyone could want, from freedom from want or fear, to any degree of decadence. The one-way cost was 500 Dracmars, which was about seven million (1871) US dollars. Human physicists warned the public that even if the ship could travel as fast as light, the one-way trip would take one hundred years, so no human passenger was likely to still be alive when the ship reached Blithe. The Questors countered that their knowledge of science was more advanced than Earth's, and they had a way of making the one-way trip in only one earth-year. This branch of science could not be shared with humans since this might disturb the natural development of human civilization. The Questors did advise prospective passengers that most beings did not plan return passages, not only because of the attractions of Blithe, but upon return to Earth, humans would find that Earth had changed a great deal, and any of their acquaintances that stayed on Earth would have died centuries earlier. Unknown to earthly physicists in 1871, this branch of science that made such a lengthy voyage possible was discovered by humans in the early 20th century; they named it "relativity."

After this bit of science fiction and alternate history, we shall begin *our* journey into *discovering* the many wonders of this exciting branch of science.

1-1 What Is "Relativity"?

All day long we scurry around, ever conscious of being on time for appointments at particular points in space, for meals, for classes, for job-related activities. Humans are creatures of time and place; in fact, consciousness itself is a manifestation of the awareness of oneself in a particular time and place. Yet in our scurryings most of us take the laws of nature for granted, particularly those associated with time and space; they seem to affect few of our daily decisions in any direct way. Nevertheless, over the centuries scientists have pondered the nature of time and of space and have explored these "givens" in our lives. The most famous and successful study of these ever-present physical realities is Albert Einstein's Theory of Relativity, which attempts to define what truths characterize time and space. Almost everyone has heard of relativity, but most have little idea of what it is. They know that Einstein developed it, so they think that it must be very difficult to understand. Only a very few have read a detailed development all the way through, and if they have, it is likely to have been a verbal description, possibly with some very abstract diagrams. Some Public TV programs and web sites can help, but they are too brief and try to get to the exciting results without starting at the beginning. Relativity is often presented as being weird. At best, the result for the audience is a hazy understanding with little knowledge that can be recalled. Dictionary definitions of the word "relativity" are usually not understandable until the reader is at least a little familiar with the subject. We will look at a dictionary definition at the end of our development, but for now we will leave the definition open. Trying to explain or understand relativity without an organized approach to the subject is simply futile. The best way to understand the meaning of relativity is to start at the beginning, go through the development, and see for ourselves. Our development *will* do just that, and hopefully will provide some real understanding.

On a practical level the word "relativity" means that two observers or experimenters that are in motion *relative* to one another will obtain certain differences in measurements of the same event or process. This does not sound too exciting, but its implications on our understanding of the universe are truly profound, and some are amazing. The theory of relativity has been with us for about a century now and it has been confirmed by many experiments, yet many people still have trouble believing some of its predictions.

1-2 Is a Physical Idea True?

Because of the skepticism in some minds about relativity, we should first briefly discuss what truth and progress in science are like, both ideally and typically. Developing a physical theory *ideally* goes something like this: Some non-understood effects are noticed or some experiments yield puzzling results. This catalog of facts hopefully leads to a hypothesis that seems to explain the facts in terms of some *simplifying general* principle(s). (A "hypothesis" is a collection of assumed truths; the individual assumed truths are called "postulates.") The hypothesis may then suggest additional experiments to test its validity. This process continues until one of two things happens: 1) The hypothesis fails to explain some effect. In this case the hypothesis is either modified or rejected. Or, 2) the hypothesis continues to explain all applicable experimental results (many experiments over an extended period of time). At this point the hypothesis is called a "theory." Since science is an ongoing process, at some later time the theory may be found to need modification or rejection. We can never be absolutely sure that a *scientific* theory is correct. That is, it cannot be proven! Regardless of past successes, a hypothesis or theory is always exposed to the possibility of being overturned by a counterexample or an incorrect prediction of some experimental result. However, as more and more experimental results are accurately explained or predicted by the theory, we have increasing confidence in it. We should have little confidence in any "theory" with little experimental support.

With all of this said, was relativity discovered along the ideal lines described above? Actually, no. It is more *typical* for scientific theories to evolve as Einstein's did. There was a famous experiment that is today cited in many textbooks on relativity called the Michelson-Morley experiment (1887) that could have been interpreted as showing that the speed of light (very important in relativity) was the same, to quite high precision, in every direction in spite of Earth's orbital motion. This could have had an influence on the development of relativity, but Einstein had not heard of this particular experiment at the time that he discovered the basics of relativity. The experiment won the Nobel Prize in physics two years after Einstein published his theory in 1905. He was familiar with the results of other experiments and also some theoretical inconsistences; he was basically trying some ideas (postulates) to try to remove these inconsistences. Still, no matter how a hypothesis or theory comes about, experimental verification is absolutely necessary in the development of scientific understanding.

1-3 Light and the Aether

Light and its speed are central to all aspects of the theory of relativity. Before we discuss light, it is helpful to consider some more familiar types of waves. Sound waves (wiggling air) move at a certain speed (about one mile in five seconds and dependent on the temperature) relative to the air (the medium) through which it is passing. Water waves move at certain speeds relative to the water (the medium). The speed of water waves is dependent on several things, including the wave frequency (wiggles per second) – tsunamis are very low frequency and move at hundreds of miles per hour compared to ripples which are at a higher frequency and move quite slowly. According to some experiments, light also seems to be a wave. (Much more about this later.) However, unlike sound waves and water waves, light can travel through the vacuum of space and at a very great speed (about 300 million meters per second, or 186,000 miles per second). **This rate is called simply, c.** This speed seems incredibly fast to us, but on an astronomical scale it is actually quite slow. Light requires about 100 thousand years to cross our galaxy, the Milky Way. Since light behaves like a wave as it travels, it seems reasonable to assume that it would need some medium. Like other waves, it would wiggle and move relative to the medium. Over a century ago this hypothetical medium was called the "aether." [The author's favorite definition of the word (from Ben Kristoffen): aether (ē′ thᵉr) n. Greek for something pure in the rarified air on Mt.Olympus breathed by the gods.] Many experiments were performed in an effort to detect this aether, especially the "aether wind" that supposedly should blow past Earth because of its high orbital speed around the sun. All attempts to detect the aether failed. Puzzling experimental results piled up awaiting explanation. (See Sec. 1-2.) Since the aether cannot be detected, it seems easy simply to assume that it does not exist. Still, the question remains, "With respect to what does light move at its characteristic speed, c?" Without the answer to this fundamental question, investigators were unable to make significant progress. They had the answer from electromagnetism, but it was unbelievable to most researchers. We will very soon discover this surprising but simple answer.

1-4 Reference Frame

Isaac Newton realized that whenever physical things are considered in any way, all locations, velocities, times, or *any other measurements must be from the point of view of a* reference frame (also a "frame of reference" or simply a "frame"). A reference frame is merely a system of coordinates with which we specify the location and time of some event(s). Since we live in three-dimensional (3-D) space, we need three space coordinates. These could be north, east, and up; or (x, y, z). Including the time the

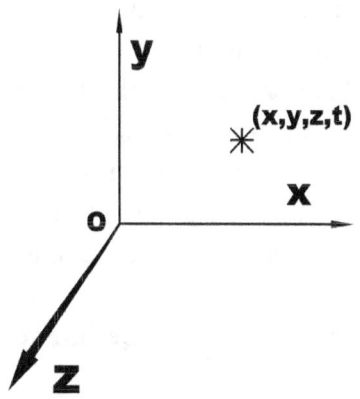

Figure 1-1 A reference frame and an event.

3

event occurred, we might use (*x*, *y*, *z*, *t*) as the location in space and time of the event. In the reference frame shown in Figure 1-1 the x-axis is pointing to the right, the y-axis is upward, and the z-axis should be seen as pointing out of the paper toward the reader, but slightly left and downward from the eyes. A negative position, for example, to the left of the origin, O, would have a negative value for *x*, etc. A **"rest frame"** is a frame in which something is at rest, i.e., not moving *in space* (the *x*-, *y*-, and *z*-coordinates are constant. Time will always be advancing.) *Our* rest frame is the frame in which *we* are not moving. The rest frame of a spaceship is the frame in which the coordinates, *x, y,* and *z,* of this spaceship, which give its position, are not changing with time.

We will essentially need only the notion of reference frames. Except in two optional appendices, we will make only small use of any of these coordinates except for time.

1-5 The Hypothesis of Special Relativity

Now that we have the basic principles of hypotheses, light waves, and reference frames, we will begin our development of the theory of relativity. To get started we need a hypothesis. This means we need ideas stated in explicit terms that we assume to be true. These statements should be consistent with experimental results, and hopefully also seem reasonable to us. If this hypothesis produces a result that in a later experiment is found to be incorrect, then we must start over. It is a tough world for hypotheses and theories.

Any organized treatment of relativity might give one the idea that the historical development was all very straightforward. However, science is not really done by a science-fair method: "Today I will discover the relationships between space and time considering that the speed of light is finite." Much work done over the last 100 years has provided us with fairly simple ways to approach the subject of relativity. The hypothesis that we will present was new in 1905. At that time there were other postulates about the workings of the universe that had been around since Isaac Newton or before. These include such things as space being "isotropic." This means that space has the same characteristics in all directions. Much evidence suggests that this is true. There may be other assumptions that will enter our discussions without being mentioned. Since experimentation is the arbiter, or test, of all physical thinking, agreement with all experiments to date is confirmation (in support, but not proof) of such assumptions.

The hypothesis of relativity will be in the form of two postulates. Although the first postulate can be stated rather succinctly in physics terms which have been carefully defined, it will require two sentences to state, and a few paragraphs to explain in layperson terms.

Postulate 1: No experiment can detect *absolute* uniform motion

("Uniform motion" means no changes in speed or direction, that is, straight line motion)**; the "no absolute" words mean that only the motion of one thing *relative* (there's the word) to something else is meaningful. The "no absolute" words also mean that there can be no basis for a preferred reference frame in which to state the laws of nature.**

This is because one reference frame is as valid as any other. Postulate 1 is demonstrated in our experience flying in an airplane. Everything will seem perfectly normal when traveling at a *constant* 500 miles per hour (or any other speed at which the plane can be flown) in the same direction (relative to the ground below). If we do not look out the airplane window and are sitting quietly, we seem to be at rest. This agrees with our experience; the flight attendant does not have to pour the drinks with the pot, can, or bottle toward the front of the plane and have the liquid rush backward toward the cup. If things seemed different flying from being parked on the ground, then we would have detected absolute uniform motion. This postulate does not seem too disturbing, but realize what we are stating in our (negative) postulate. *The concept of _absolute_ uniform motion does not exist.* We cannot say that we (or they) are moving, only that one thing moves *relative* to the other, and that this principle holds no matter how fast the motion. In 1905 our experience did not extend to very great speeds, say, 90% the speed of light (0.9 *c*), so this postulate was an extrapolation (extension) of experience (i.e., experiments).

That there can be no basis for a preferred reference frame means that the laws of nature can be stated using exactly the same words, in exactly the same order, in any reference frame, no matter its (uniform) motion relative to another frame. If a law of nature requires some mathematical equations in order to describe it, then these equations will also be exactly the same from one frame to another. If the laws of nature did change from one frame to another, then since the only difference there can be between frames is their speed and direction of motion, then we would have been able to detect absolute motion.

It should be made clear that while there can be no preferred reference frame in which to *state the laws of nature*, there can be preferred reference frames in which to *solve a problem* or *consider a situation*. This is for convenience only; the results should have exactly the same *interpretation* using any frame. Consider the pouring of the coffee in the airplane mentioned above. We can describe the motion of the coffee between the pot and the cup in the rest frame of the airplane quite simply: The coffee leaves the spout of the pot with zero vertical velocity, then acquires a downward velocity as it falls under gravity. Since the cup is directly below the spout, the coffee falls into the cup. To describe this same process in the rest frame of Earth's surface, we must also consider some other laws of nature that we did not need to use in the plane's rest frame. The description is also more involved: The coffee is moving, say, west at 500 miles per hour as it leaves the spout of the pot. Because no horizontal forces are acting on the liquid in transit, it maintains its horizontal speed of 500 miles per hour to the west as it falls. Its vertical speed does increase downward as the coffee falls toward the cup that is also moving westward at 500 miles per hour and therefore maintains its relative position directly below the spout. The reader need not completely understand this second description at this point. We are only making the point that considering a situation in one reference frame may be simpler than another. At most places in this study of relativity we will select the most convenient frame of reference in which to describe that situation. We do not need to make the situation more complicated than need be. This first postulate is usually called "The Principle of Relativity." Not too shocking, yet!

Before leaving the discussion of Postulate 1, we should address a difficulty with these concepts that readers usually experience at this point. Postulate 1 says that any frame of reference is as legitimate as any other if it is moving at a constant speed and direction. Thus, any observer may always consider herself to be at rest in her own reference frame. If you are in a closed up airplane flying smoothly, and you do not look out the window, it is reasonably easy to believe that you are at rest, and things within really do not seem any different from being parked on the ground. It simply looks and feels this way. However, if you are driving down the highway and see the scenery passing, with its trees, buildings, and billboards, it is difficult, possibly impossible, to envision this input data to your brain as your car being at rest, and the scenery dutifully passing by you. There are several reasons for this difficulty: The scenery is big, and you in your car are small. Your ride is not perfectly smooth. You also hear sounds that we attribute to motion, such as engine, wind, and tire noise. We also (think we) know that the car is taking us from one place to another, so we (think we) know that the car must be in motion. Also, after years of driving practice, it would be difficult to believe that we had to turn the steering wheel clockwise to make the scenery move to the left. This would be the same even in countries which drive in the left lane.

Nonetheless, Postulate 1 states that we *may consider* the car and ourselves to be at rest, and everything outside the car to be in motion relative to the car. The laws of nature will not be different in doing so, as long as the reference frame moves at a constant speed and direction.

Some video game players may have an easier time with this. In many video games the person playing the game takes the form of some entity that moves through some elaborate structure destroying some things and gathering others. The player's alter ego is usually fairly stationary on the screen, the digital environment doing almost all of the on-screen motion. The player feels no motion in all of this, but the joystick, mouse or keys still seem to control the player's motion, rather than the structure's.

Many times in our development of relativity, the reader will be asked to imagine himself to be at rest in a particular reference frame. Other observers, clocks, and even spaceships will be claimed to be moving in this frame. It might help if you envision yourself as being the large, immovable, omnipotent ruler of this corner of the universe. You have chosen a region of empty space where various types of objects will be passing by, performing experiments to give you pleasure. So that you may obtain a different viewpoint, you may be asked to consider yourself to be moving along with some spaceship, clock, or particle. Alternately, trusted observers who are moving along with the experiment will report their observations to you so that you may discover truths and write your treatise on the workings of the universe. As mentioned earlier, relativity usually involves viewing experiments from two different reference frames. We will see this shortly in Figures 1-3 and 1-4. This book is a guide to save you from having to figure it out by all by yourself. To benefit from reading this book, you should try to follow each step in the logical arguments. After this lengthy discussion of Postulate 1, we will now proceed to our second postulate.

Postulate 2: The speed of light is not affected by the motion of its source. "Source" includes surfaces off which light has reflected.

This postulate is not so familiar to us as is Postulate 1, but then how many times have you measured the speed of the light emitted by a moving light bulb? However, astronomical observations strongly suggest that this postulate is correct. For example, most stars are in multiple star systems. Our sun is an exception, being a single star. In some of these multiple star systems, the stars are moving in orbits at very high speeds (relative to each other and usually to Earth). These speeds can be hundreds of miles per second.

Consider the left binary star system (two stars) shown in Figure 1-2. There are many such systems, so it is common for Earth to be fairly close to the orbital plane (the flat page in Figure 1-2) of the stars in some of these systems. Thus, usually the stars will be moving, at least partially, toward or away from us. If the stars were not orbiting at high speeds, then gravity would quickly pull them

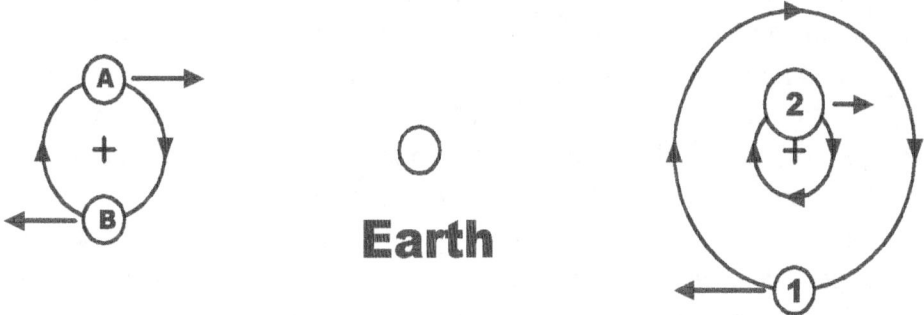

Figure 1-2. Two binary star systems are shown grossly out of scale. Earth would be much smaller and many times farther away from each system. The orbits (paths) of the stars are shown by the circles with the pointers. The stars are of equal mass in the system at left, but star 2 is more massive than star 1 in the system at right. The orbital speeds are symbolized by the lengths of the arrows.

together. At any one moment they must be moving in opposite directions because the stars must form a seesaw that is balanced at the common center of the two orbits (the "+" sign). These stars can be so far away from Earth that it takes light, even at its great speed, hundreds or even thousands of years to reach Earth. (A "light-year" is the *distance* that light will travel in one calendar year. This is about six trillion miles, or about ¼ the distance to the star nearest to our solar system.) If the motion of a star affected the speed of its emitted light, then the travel times for light from this system to reach Earth would be different for the two stars; perhaps a lot different. If the travel times were different, observers on Earth would not see the two stars where they were located at the same moment in the past. The star emitting the faster light would reach Earth before light from the other star. Since we *see* where something was by using the light that is arriving at the moment, one star would be seen farther back in time than the other. For example, if the pair of stars is about 100 light years away and motion *did* cause a change in the emitted light speeds, then we might see one of the stars where it was 100.000 years ago and the other where it was 100.001 years ago. That is about an 8-hour difference. In eight hours, stars in these close, high-speed systems can move a big fraction of the way around their orbits, so the star system would not look normal to us at all. It would appear to be violating the laws of physics. The two stars might not appear on opposite sides of their orbit, as they must be. At some moment observers on Earth might even see one of the stars at more than one place, or not at all. Stars that orbit at high speeds are quite close together. The time required for light to traverse this small separation might be less than one minute, so this is a very small effect compared with the potential effect of differences in the speeds of light combined with the hundreds of years that the light must travel to reach earthly observers.

The remainder of this section covers a relatively unimportant point. It is included for those readers who wish to see the complete supporting argument for Postulate 2. It is somewhat tedious. The real justification of each postulate is that together they predict the correct outcomes of experiments. Unless you are a stickler for detail, you may want to skip the next two paragraphs.

In the discussion above the reader probably assumed that *if* the motion of the star *did* change the speed of its emitted light, then the light's speed would be increased in the direction of motion of the star. This would certainly be our first guess. However, in the argument above we made no such assumption, so the argument holds even for the opposite guess. (The speed of emitted light *decreases* in the direction of motion of the source – not true either.)

Now suppose that the speed of the emitted light is simply affected by the star's speed and that there is no dependence on the *direction* of the star's motion. In such an unlikely case our argument above is incomplete! However, our binary-star evidence still strongly suggests Postulate 2 because usually the two stars in these systems are not the same size; that is, there is not the same amount of stuff (mass) in them. In this situation the two stars must be in different size orbits with the *larger star* in the *smaller orbit;* the seesaw must balance. This is shown in the system to the right of Earth in Figure 1-2. The two stars orbit in the same amount of time, so the smaller star, completing its larger orbit in the same amount of time, will have the greater orbital speed. These unequal binary star systems also appear to follow the same laws of motion regardless of their distance from Earth, whatever the angle of their orbital plane as we see it, even after their light has traveled for hundreds of years. This evidence strongly suggests that speed alone (no directional dependence) also does not affect the speed of emitted light.

It could be argued that a star's motion does cause an effect on the speed of the emitted light, but that it is far too small an effect to be detected with our equipment. However, we extrapolate again and postulate that there is no such effect, regardless of the relative (the only meaningful) speed between the light source and the observer or anything else. Again, the justification of this assumption will be if relativity agrees with all experiments and observations.

For example, Astronomers detect atoms emitting and absorbing light and radio waves from all over the universe. These distant atoms seem to be identical to, and behave the same way as earthly atoms. This gives strong evidence that the same laws of nature are at work all over the universe. Light and radio waves travel at a finite speed, *c*, so when we look deeply into space we are also looking far backward in time. Again, indicating that the universe is not too outlandish, the actions of the atoms seen from long ago and far away seem to be identical to those we have here and now. With this evidence in hand, we will assume that our postulates are true throughout the universe and have always been. In Chapter 5 we will add a third postulate when we look into gravity. Until then, all that is left for us is to discover where these two postulates lead us. If we ever arrive at a conclusion that is counter to experiment, then we must modify our postulates. We will be doing most of our logic with **"gedanken experiments"** (German for "thought experiments"). Parables are not usually much help with relativity because we do not have anecdotal experience with really fast speeds, and it is at rapid speeds where most of the relativistic effects appear in any noticeable way. However, we will use an analogy about a bicycle race in Sec. 3-3 and other analogies to aid in some arguments.

1-6 Our First Applications of the Postulates

Suppose that Bob, who is standing beside a road, detects that a car moves past him at a speed of exactly 60 miles per hour moving toward the north. Ava (in the car) will see Bob as moving past her toward the south. Does she detect him moving south at *exactly* 60 miles per hour as we would assume? We must base our reasoning on our hypothesis: If she does *not* detect him to be moving at exactly 60 mph, then this would be a violation of Postulate 1 because this could form a basis to prefer one of the reference frames (the rest frame of Ava or Bob) over the other. We could have a basis to prefer the frame which produces the slower (or the faster) speed of the other observer. Thus, the first postulate implies that two observers will each determine that the other is moving at the same speed, but in the opposite direction. We will call this simple, but very important conclusion **"Corollary 1"** because it follows so quickly from Postulate 1. [We can have no basis for preferring a reference frame that moves in a certain direction, since we have also postulated that space is isotropic (the same in all directions). We do not consider the assumption of space being isotropic to be a *relativity* postulate because this had been assumed for centuries before relativity.] We will make much use of Corollary 1 in our development of relativity. Fortunately, Corollary 1 is very simple and is exactly what we would have assumed to be true.

Our next gedanken (thought) experiment produces our first surprise, which is the answer to the question, "Relative to what does light move at its characteristic speed, *c*?" Consider two rocket ships and a light source. Ship A is at rest relative to the light source (perhaps the ship and source are connected by a stick), while Ship B is moving toward the light source. See Figures 1-3 and 1-4. These show the situations as determined by Ava in Ship A and then by Bob in Ship B at the instant that the two spaceships pass each other. Each ship is equipped to measure the speed of the light that passes by it. The details of measuring speeds are in Section 1-7.

Ava will measure c = 186,000 miles per second = 300 million meters per second for the speed of the light that passes by. This will simply establish the value of c. Since any observer may always be considered to be at rest, then Bob will see the situation as shown in Figure 1-4. As

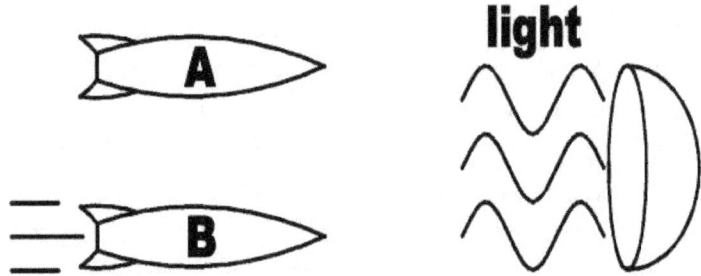

Figure 1-3 **As observed by Ava** (in her rest frame)**:** Remember, the observer may always be considered to be at rest.

determined by Bob, Ship A is backing up, and the light source is approaching. However, according to Postulate 2, the motion of the light source does NOT affect the speed of its emitted light, so Bob will get the same result for the speed of the light that Ava did, namely c = 186,000 miles per second.

From Postulates 1 & 2, this gedanken experiment has established the shockingly simple conclusion that **LIGHT MOVES AT c RELATIVE TO THE OBSERVER!** Every being in the universe will measure light from whatever source as moving relative to himself at this same speed, c. Since we arrived at this conclusion without having to specify the speed of the rockets (observers), then this result must be true for any speed of the observer. So, "Relative to what does light move at its characteristic speed?" Answer: Whoever, or whatever, is making the speed measurement. Since any observer could, in principle, measure the speed of light in his own frame, then light must move

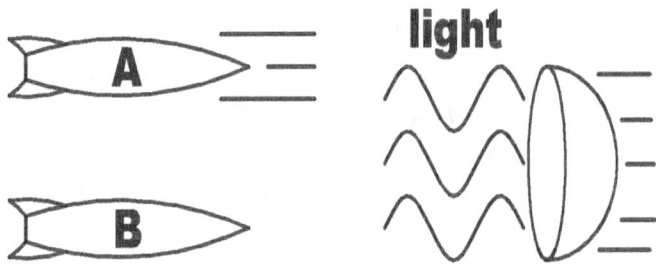

Figure 1-4. The same situation **as observed by Bob** (in Bob's rest frame). Ava and light source both moving to the left relative to Bob.

at c relative to any observer even if he has not measured it. Since we have shown this to be true with the preceding gedanken experiment, any observer can have confidence in it. We will call this surprising result **"Theorem 1."** We assume that space is isotropic (the same in all directions, Section 1-5), so this result holds in all three space dimensions. This makes sense because we could merely turn Figures 1-3 and 1-4 in other directions and make the same arguments. We will make great use of Theorem 1 throughout our development of relativity. Do not confuse Theorem 1 with Postulate 1. We will have limited future need for Postulate 2 even though it was essential for arriving at Theorem 1. Thus, Postulate 1, Corollary 1 and Theorem 1 will form almost all of the basis of our remaining logical arguments.

But you may say, "I've never heard of anything so ridiculous! How can it possibly be the case that different observers that are moving with respect to each other and even moving parallel to the light beam will obtain the same measurement for the speed of the *same light* that happens to pass by them? Surely, if I am running from the police and they shine a flashlight toward me, the light will pass by me more slowly than if I am running toward the flashlight. I know that this is true for bullets that they fire at me." Remember, the speed of light is so fast to us that our everyday experience has not equipped us with knowledge of what happens at speeds in this realm, so we must proceed carefully. Still, until we see how well Theorem 1 works in describing the universe, we must content ourselves with the fact that this is in complete agreement with experiments which measure the speed of light.

Can we explain how obtaining the same measurement for the speed of light comes about? Corollary 1 rules out one potential source of explanations; both Ava and Bob will each detect that the other is moving at the same speed (but in opposite directions) relative to each other, so different relative speeds of the observers cannot become the basis for an argument. Also, we could reverse the direction of ship B relative to ship A, and the argument would not be changed. The only remaining possibility is that motion must somehow cause changes in the speed-measuring apparatus. Since speed is a distance divided by a time interval (for example, miles divided by seconds, written as miles/second, see Section 1-7), then perhaps motion makes changes in distance or time or both (but in ways that always produce a constant speed of light). This turns out to be the correct explanation as we will show as we further develop our knowledge of relativity. We will eventually get used to Theorem 1, and then it will no longer bother us. Fortunately, Theorem 1 is also the simplest possible result and the easiest to use. This amazing result greatly simplifies the analyses that we will apply to most of the gedanken experiments in our development of relativity.

1-7 How Can We Measure a Speed?

We can determine speed from the old equation: Distance, D, equals rate (speed or velocity, v) times time. That is, $D = v\,t$ or $v = D\,/\,t$ or $t = D\,/\,v$. These last two forms are: Speed = distance over time, and Time = distance over speed. (Consult Appendices A1-1 and A1-2 if you are not sure how these last two equations come from the first equation.) In principle, the gedanken experiment shown in Figure 1-5 will work to measure the speed of light. Let us place photodetectors on opposite ends of a meter stick. A stop clock is located at the middle of the meter stick and is connected to the photodetectors via equal lengths of wires. A light pulse from a flash passes parallel to the stick.

When the light pulse reaches the *first* detector the clock *starts* after the electrical signal travels through the wire to the clock. Light reaching the *second* detector will cause the clock to *stop* after

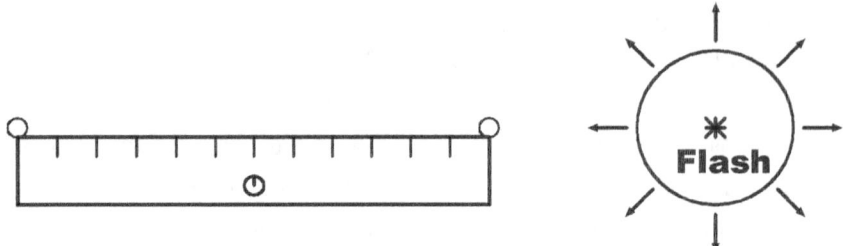

Figure 1-5. Measuring the speed of a light pulse: Photodetectors (shown as small circles) are connected via equal-length wires (not shown) to the stop clock located at the middle of the meter stick.

that electrical signal reaches it. We then use $c = (1 \text{ meter}) / t$, where t is the elapsed time as measured by the clock, to calculate the speed of light. The time, t, will be only about 3.33 billionths of a second, but some electronic clocks are up to the task of measuring such short times.

We could use the same method to measure the speed of anything else, a thrown baseball, for example. We would only need to replace the photodetectors with baseball detectors. If we need a method to measure the speed of a passing spaceship, we will use passing-spaceship detectors. For the remainder of our study, we will assume that such a straightforward method will be used to measure all speeds, or a slight modification using a pair of synchronized clocks, one at each detector, where we subtract their readings to obtain the time interval.

1-8 Time, Distance and Other Measurements

Throughout our development of relativity we must be quite explicit about how we will make measurements. We might make mistakes if we simply say something like "measure the length of the approaching stick" Also, we must apply what we have already established about relativity to the logic at each point in our analysis of the gedanken experiment. For the moment, the reader will have to trust that our development is including all of the necessary considerations. After completing this study, then the reader will be able to go back and verify most of this.

What is time? We think that we know, but when we try to say what it is, it eludes us. It has been said that "Time is what keeps everything from happening all at once." Humorous, but not very helpful. Or, "Time is money." This is economics or business, not science. We will adopt an *operational* definition of time: TIME IS THAT WHICH IS MEASURED BY CLOCKS. This may seem like double talk or Newspeak, but it is the only definition that works, and we will see that it works very well. You may ask, "OK, but what is a clock?" The most common type of clock is a device that counts repetitive events. Probably the simplest clock might be called the "castaway clock." This is some surface upon which some survivor places a mark for each passing day (sunrise or sunset). A calendar is a clock only when something, or someone, marks off the days based on the position of the sun in the sky. To measure time in smaller units than days we must use something with a cycle time that is shorter than a day. Hourglasses were used, but it was found that more sophisticated devices, such as pendula with mechanisms to maintain and count their swings, were more consistent with the length of the day. We now use electronic devices to maintain and count the vibrations of quartz crystals, or even atomic clocks that count frequencies that are emitted or absorbed by atoms. These last two types of clocks are much less affected by disturbances such as vibration or temperature change when compared with older types of clocks.

Distance is similarly defined: Distance is that which is measured by meter sticks. We would have to include instructions on the use of the meter sticks. Actually, the definition of every physical concept relies on an operational definition, that is, a description of how it is measured.

To be confident that we know what we are doing with operational definitions, a variety of measurement methods, or locations, or whatever must be consistent with each other. Otherwise, the concept that we think that we are measuring, say electric charge, might not make any physical sense.

1-9 The Transverse Light Clock

Now, another gedanken experiment: This experiment will produce the quintessential result that we must use in our understanding and development of relativity. Many people have heard of this result, but they may be surprised at how quickly it is derived. Make sure that you (at least, qualitatively) understand what is being developed in this section. Imagine that *they* (observers that are moving relative to *us* in *our* rest frame) have two mirrors at rest in *their* frame and spaced a distance L apart (as measured by *their* observer in *their* frame who is at rest relative to the mirrors. Being at rest might make a difference. We will later see that it may, depending on the direction of the observer's motion.) See Figure 1-6. A very brief flash goes off that causes a pulse of light to bounce up and down between the mirrors. Sensors detect each bounce and produce "ticks" when the light pulse strikes

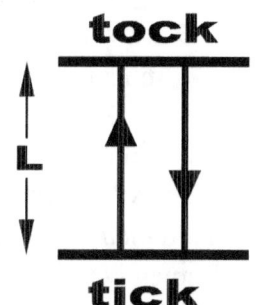

Figure 1-6. A light clock as seen by an observer at rest relative to the clock.

the bottom mirror and "tocks" at the top mirror. The time for one complete cycle, tick tock tick, will be $t' = 2 L / c$ as determined by *them*, who are at rest relative to the mirrors because time equals distance over speed, and the distance is up plus down = $2L$. For some reason the " ′ " symbol is called "prime." Time should usually be measured by counting *complete* cycles of something. For example, the time between summer and winter on Earth is not exactly the same as from winter to summer because of the slightly elliptical shape of Earth's orbit around the sun, and the resulting changes in Earth's orbital speed as the year progresses. The tilt of Earth's axis also affects this.

If we took one of these seasonal times and simply doubled it, we would calculate the length of the year slightly incorrectly. However, if we can be really sure that the half cycles of whatever we are counting are of equal time, then we may double. You may ask, "How would a light clock display its reading?" To complete the clock, electronics would be needed to count the ticks and the tocks. When enough ticks and tocks have been counted so that the light pulse has traveled a distance up and

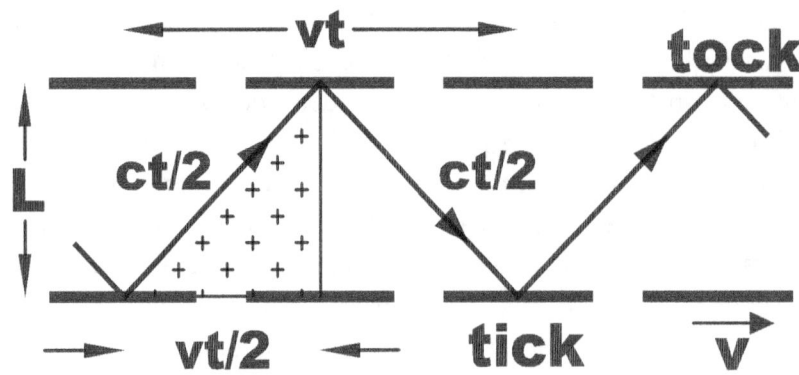

Figure 1-7. The Transverse Light Clock as seen in **our rest frame**. The light clock is moving to the right with a speed v. The two mirrors are shown at their positions each time the light pulse reflects off either mirror.

down of 300 million meters (= 186,000 miles), then the clock's display will advance one second. In this gedanken experiment *we*, in our frame, will be moving to the *left* at a speed v relative to the mirrors and to *them*. Thus, *to us*, the mirrors (the light clock) will be moving to the *right* at the *same speed*, v. Remember Corollary 1 in Section 1-6?

     [Vitally Important! Throughout our entire study of relativity, when we draw any diagrams, indicate or make any measurements, pose any questions, offer any answers, or do any calculations, they must always be with respect (relative) to <u>one specified</u> reference frame!] We will see many examples of this as we proceed. Refer back to Figure 1-6. As seen by observers at rest relative to the mirrors, that is, *them* (specifying the reference frame), the light pulse simply bounces up and down between the mirrors. The light paths are shown displaced sideways for better visibility in the figure, but actually follow the same path up and down (in this reference frame).

     Figure 1-7 shows *their* light clock **as seen by *us*** (in our rest frame) at four sequential times when the light pulse reflects off a mirror. In our frame the light path is a zigzag line, and Figure 1-7 covers slightly more time than two cycles of the clock (tick tock tick tock).

     To us, they, and the mirrors, are moving at a speed v to the right [*transverse* to the separation of the mirrors ("Transverse" means perpendicular, or at right angles to. In this case the separation of the mirrors and the direction of the light in the rest frame of the mirrors will be perpendicular to the relative motion between the two observers.)] The time between ticks *to us* will be called t. (It is t'

to them.) We do not yet know how long t will be; nevertheless, we give it a name so that we can more easily work with it.

Imagine that there is a little smoke between the mirrors so that we can see the light paths; think of a laser light show. The term "smoke and mirrors" usually refers to magic shows, not to our gedanken experiments, but we will use it anyway. In this (our) frame the light pulse still makes equal-angle reflections off the mirrors (90 ° in their frame, but a smaller angle in our frame), but to us the light pulse has to travel an angled distance that is greater than L between tick and tock, and also between tock and tick. Since by Theorem 1 light **moves at c relative to the observer, *us*, the longer light path will mean that there is a longer time interval** to us between ticks and tocks than *they* will determine. Thus, *we* **will determine that *their* light clock (moving with them) runs slow(ly)!** If the speed, v, is even greater then the path of the bouncing light will be even closer to horizontal than shown in Figure 1-7 (with the lesser speed). This will mean an even longer time between ticks and tocks to us. (Remember, in a vacuum, light always travels at c.)

How much slower would their light clock seem to us? Some math provides the answer. The path of the light going down is clearly equal in length to the light path going up, so we are going to calculate the tick-tock time and call it half the cycle. Make sure that you notice the /2's in the labels of the (shaded) right triangle (one angle is 90°) in Figure 1-7. During the time $t/2$, light will travel a diagonal distance of c times $t/2$ (distance = rate x time), and the mirrors will travel a distance of v times $t/2$ to the right. Using the Pythagorean Theorem [The square of the hypotenuse (the side opposite the right angle) is equal to the sum of the squares of the other two sides. (See Appendix A4-1)], **we obtain** (restating which reference frame is being used for the calculation)

$$\left(\frac{ct}{2}\right)^2 = \left(\frac{vt}{2}\right)^2 + L^2 \qquad (1\text{-}1)$$

This equation can be solved for t in terms of L, c and v. Then remembering that *to them* the time between ticks is $t' = 2L/c$, we can manipulate Equation 1-1 (See Appendix 6), to obtain:

$$t' = t\sqrt{1 - v^2/c^2} \qquad (1\text{-}2)$$

We will be seeing this particular square root, $\sqrt{(1 - v^2/c^2)}$, many more times. See Appendix A3-1 for an explanation of squaring and square roots. Later we will discover that only speeds, v, less than or equal to c, are possible ($v \le c$). This square root will have values between zero and one. See Table 1-1 in Section 1-10. Thus, for any speed other than zero, from the point of view of the *un-primed* frame (us), t' (their time between ticks) will be less than t (our time between ticks).

Most important: All of this is as determined in *our* frame about the goings-on in a frame (their frame) that is moving past us at a speed v. Both from Figure 1-7 with the longer light paths than in Figure 1-6, and also since the square root is less than 1, we conclude that the moving light clock runs slow. This means that in our frame, we conclude that when some amount of time, t, has

15

passed (to us), then less time, t', will have passed to them. **That is, to us, their *clock* slows down. Then by definition** (Sec.1-8)**, their *time* slows down to us (the outside observer).**

(This paragraph is optional. We will not use its conclusions in our development.) Another way of interpreting Equation 1-2 exists: If we consider two events composed of the light pulse reflecting off, say, the lower mirror, then these events occur at the same location in space in the primed frame (their frame, in which the mirrors are *not* moving). However, in *our* frame in which the mirrors *are* moving at a speed v, these two events are separated by a distance, vt. *For these two events*, we can see that t' in Equation 1-2 is the time between these two events where the two events occur at the same place, while t is the time between these same two events in *our* (the observer's) frame that is moving at a speed v relative to the light clock, so the events are *not* at the same place in our frame; they are separated by a distance vt. We could try various speeds, v, and directions of motion (keeping the mirror spacing perpendicular to the motion), until the points in space where the light bounces off the lower light clock mirror (ticks) happens to coincide with any two events of our choosing. Thus, the interpretation of the meaning of t' and t earlier in this paragraph can apply to any two events.

What would *they* say about our clocks and thus about our time? To answer, we could exchange the labels "us" and "them." t' would then be *our* time between ticks, t would be *theirs*. Thus they would say that our clock runs slow. The situation is symmetrical. More convincingly, from Figure 1-7, they would see our light clock moving to the *left* at a speed v. The light path would be similar to Figure 1-7, but reversed right to left, so they would see the longer light paths for us. **Thus, we each see the clock at rest in the other frame as running slow.** This sounds crazy at first, but any outcome such as "We see them as slow, and they see us as fast," would violate Postulate 1. If we were on vacation, we might prefer the frame with the slower time. This result takes some time to get used to, but you will get used to it, and it will then seem more natural to you.

If time were to slow down within our frame, that is, if everything, even thought processes, slowed down by the same amount, then we could not tell. Think about it. Everything would seem normal to us. This is in agreement with Postulate 1. However, what would they say about their own light clock and time in general within their own frame? To them everything at rest within their own frame will seem perfectly normal, as it must be to be in agreement with Postulate 1.

Exercise for the Reader: "OK, it looks like we are forced to these conclusions about light clocks, but what about other clocks?" This exercise is to devise a simple argument, based on possible violations of Postulate 1, to show that all clocks, chemical reactions, radioactive decays, and even thought processes must slow down by the same factor. (We have not covered nearly enough information to be able to explain what would make another particular kind of clock run slow. In fact, this can be quite difficult.) Answer: "If different types of clocks in our spaceship that previously agreed with each other were now (after blastoff) found by us to be running differently from each other, then we would have detected _____ _____ _____." (Hint: See Postulate 1.) The answer is in Appendix 18.

Before we leave this section it should be emphasized that the motion that we have been considering is "linear," that is, straight-line motion. Figure 1-7 is *not* like a view from the center of a *rotating* carousel of a horse near its edge. This *linear motion at a constant speed* of one frame relative to another frame is what is meant by the word "Special" in Special Relativity as in the name of Sec. 1-5. ("Special" means a special case, rather than a general case.) More on this in Chapter 5.

This paragraph is very important: Mistakes can happen when using Equation 1-2. We must be careful to keep firmly in mind which frame of reference we are using for our logic; this frame must be the one in which we are at rest, and other frames must be considered to be moving relative to it (and to us). The following mental crutch will help to prevent confusion and mistakes: When some time, t, passes in our rest frame, then we conclude that less time (by multiplying t by that square root) will pass in any other moving frame. **We cannot conclude what they (in another frame) will determine about the goings-on in our frame without transporting ourselves (at least mentally) to their frame and <u>redoing the analysis</u>!** For example, we must not simply solve Equation 1-2 for our time, t, in terms of their time, t', because Equation 1-2 is valid only for observations made from *our* rest frame (because that is where we assumed the observer was located when we derived Equation 1-2). Remember, we each determine that the other frame has slow time, and we will discover other similar things. However, the other things, such as lengths, can be drawn, so we will be less error-prone with them than with time.

Note: The algebraic manipulations that are only math logic, but not part of the relativity logic, are presented in appendices for those who wish to see them. They are placed there because their inclusion in the chapters might disrupt the flow of the relativity logic being pursued at a critical point. Also, some readers may not want to bother with the math manipulations, and simply accept that they have been properly done. Basic math lessons are also included in the first sections of the appendices. Answers to most of the exercise questions and some additional gedanken experiments on slightly more advanced optional topics are also in the appendices.

1-10 The Relativity Factor

The square root that appeared in Section 1-9 will occur in many of our future expressions as either a multiplier or a divisor.

Table 1-1 has some sample values of this square root $R = \sqrt{(1 - v^2/c^2)}$ and also its reciprocal $1/R = 1/\sqrt{(1 - v^2/c^2)}$…for various values of v to give the reader an idea of the magnitude of relativistic effects. The reciprocal form will be needed later.

This square root will be very important in our development of relativity. Make sure that it is familiar to you. Write it down a couple of times. From here on, we will often abbreviate it. "R" stands for "the Relativity factor."

$$R \equiv \sqrt{1 - v^2/c^2}$$

The triple bar symbol means "is defined as" or "is identical to," not just equal to under some condition (like in a "word" or "story" problem or gedanken experiment). We will still make considerable use of the square root expression. In many other treatments of relativity, especially those that deal mostly with physics, then the symbol γ (lowercase gamma) is used for the reciprocal of what we have called R. That is: $\gamma = 1 / R$. In physics equations the reciprocal of R appears more frequently than does R. Notice that if $v > c$ (v greater than c, 1.5 c in Table 1-1) then each of the right two columns becomes an imaginary number (see Appendix A3-2) and the right column even becomes negative! It is difficult to make physical sense of this, so we will tentatively restrict our thinking to $v < c$ (v less than c) for things other than light. Later we will discover better reasons to do this.

Table 1-1. Sample values of *R*, the Relativity Factor (the square root), and 1/*R*:

Speed, v	$\sqrt{1-v^2/c^2}$	$1/\sqrt{1-v^2/c^2}$	
0	1	1	(No relativistic effects)
$c/100$	0.99995	1.00005	(1860 miles per second!!)
$c/2$	0.86603	1.15470	
$3c/5$ (0.6c)	4/5 (0.8)	5/4 (1.25)	(from 3, 4, 5 right triangle,
$4c/5$ (0.8c)	3/5 (0.6)	5/3 (1.67)	see Appendix A4-1)
0.9c	0.43589	2.29416	
0.99c	0.14107	7.08881	
0.999c	0.04471	22.36627	
0.9999998c	0.000632	1581	(the muons in Sec. 1-16)
c	0	∞	
1.5 c	1.11803 i	$-0.89443\,i$	(v is faster than light)

1-11 More about Light and c

Why is light so special that it so frequently enters these discussions? It is not really light itself, but its speed that is important in our considerations. The speed, c, is a constant of nature. It is numerically equal to the speed of things that can only exist if they travel at this speed (in a vacuum). More things than light move at c. Gravity and electricity also travel at this speed. (Neutrinos were once thought to as well, but it has been found that they actually do possess a very small mass, and so can only travel very, very near c, but not quite at c.) Except for neutrinos, these are all called "massless particles" because if you stop them you have nothing left but the energy and momentum that they gave to something else while stopping. (Energy and momentum will be described in Chapter 4.)

Light comes in lumps or "quanta" called "photons." It helps to think of a photon as a clump of waves, sometimes called a "wave packet." The confusion that existed for centuries about whether light is waves or particles is thus answered. It is both. When light is emitted or absorbed it behaves more like a particle (Einstein won his Nobel Prize for this, not for relativity.), but when it travels it is

Figure 1-8. A plot of either the electric or the magnetic field of a photon. There would actually be many more oscillations.

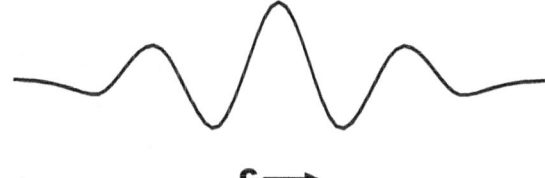

c⟶

more like a wave. Light waves are not displacements of some medium as in sound or water waves, but rather are electromagnetic waves (oscillating electric and magnetic fields). The electric and the magnetic fields are perpendicular to each other and also perpendicular to the direction of travel. More on this in Section 5-1.

The primary reason that we have used light clocks in our treatment of relativity is that this is the only kind of clock that, so far in this study, we have the resources to treat. Any other type of clock, such as a quartz clock, involves much more complex calculations. For most kinds of clocks, it is quite a difficult analysis. We will not attempt such calculations in this book.

Another reason that we deal with light is that it is familiar to us, and we can see it. Maybe some of our development of relativity will involve parables after all.

1-12 Transverse Lengths – Cleaning Up a Loose Detail

You might say, "Hey! I've got an idea. I think that I see a problem. Maybe we can be saved from this craziness! I've heard that things shrink when they move really fast. In the Transverse Light Clock analysis we assumed that the separation between the mirrors was the same in both frames, namely, *L*."

If you also thought of this, then give yourself a gold star. We did make the assumption that the separation between the mirrors was *the same* in both frames. This must now be proven to be correct. In our proof we must not make use of the slowing ("dilation," or stretching out) of time that we just discovered, because part of the argument (that went unstated) was that the measured separation between the mirrors *was* the same in each frame. We can provide the proof we need with the following gedanken experiment:

Notice that the mirrors in the transverse light clock were separated transversely (perpendicularly) to the line of motion of the two frames. Let us consider two meter sticks as shown in Figure 1-9. One stick has pins stuck through it. The sticks move relative to one another with the sticks perpendicular to the motion. As they pass, the pins leave scratches on the plain stick. We can then stop the stick(s) and bring them back side-by-side at rest *relative* to one another. (There's that word again.) Suppose for argument that motion does shorten a stick. As determined in the rest frame of the plain stick, the stick with pins *is* moving. If motion caused the moving stick to shorten, then the pins would not match the scratches when the sticks were reunited at rest relative to each other. After laying the sticks side-by-side, the scratches would be closer together than the pins because the stick with the pins was supposedly shortened when it was moving. If we then consider the same experiment from the rest frame of the stick with the pins, then the plain stick would be moving and thus be the shortened one, and its shortening would cause the scratches to be more spread out than the pins when the sticks were laid side-by-side at rest. Think this through. If the experimental results were not identical when we reunited the sticks, we could then, *based on assumed shortening*, determine from the placement of the pins and their scratches which stick was moving (rather than only relative motion). This would be a violation of the First Postulate in the First Degree. The penalty is 10 years of imprisonment, as measured on the warden's watch. Since the warden moves around so fast, his watch seems slow to you, locked in your cell. Your attorney might argue that motion lengthens sticks, but that would not alter the essence of the argument nor the conclusion: **Transverse lengths are not altered by motion.** Not everything is relative, that is, different in different frames; some things are absolute. Examples of absolutes we have found so far: the speed of light, and now, transverse lengths.

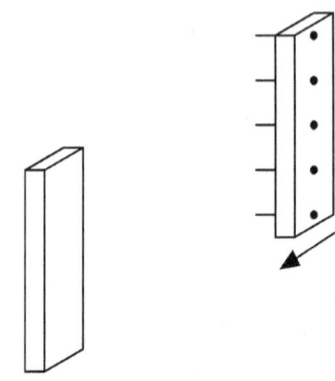

Figure 1-9. Two meter sticks. One has pins. Shown in the rest frame of the plain stick.

1-13 Length Contraction   (also known as the "Lorentz Contraction")

We are now in a position to easily discover another relativistic effect, an effect that you may have heard of. Earlier, we asked the question about how it is possible for every observer to obtain the same value for the speed of light, even if she were moving toward or away from the light source. We have previously discovered that, *as we determine it* (in our frame), a moving observer's clock(s) (at rest in her frame and her time) will be running slow by the factor $R = \sqrt{(1 - v^2/c^2)}$. Since speeds are found by distance (length) / time, then since time intervals are reduced (slowed) in a moving frame by this factor R, lengths must also be reduced by the same factor to always get the same answer for the speed of light, namely c. (If you need help on this see Appendix A1-2.) Looking back at our gedanken experiment about measuring speeds (Section 1-7), and remembering the result of Section 1-12 that transverse lengths are not affected by motion, then it must be that it is **only lengths parallel to the motion that are contracted.** It does not matter which sense (to the right or left) the motion of the meter stick with the detectors and the clock in Figure 1-5 has. This is indicated by the symmetry in the gedanken experiment shown in Figure 1-9 as well as v^2 in the relativity factor; positive or negative speeds will give the same result when squared. (See Appendix A3-1.)

If we speed past another observer, or he by us (exchanging rest frames), he would say that all of our "longitudinal lengths" (lengths parallel to the motion) were contracted, and we would say the same about all his longitudinal lengths and by the same factor (Corollary 1). As it is for time, if *all* our lengths in one direction (or even in several directions) were shortened by the same factor, we could not tell. Our measuring tapes and eyes would shrink by the same factor as everything else. This similarity to time slowing down with motion gives a certain symmetry to nature. All of this is in agreement with Postulate 1 (We cannot detect absolute uniform motion.), as it must be if Postulate 1 is correct and we have not made a logical error to this point.

Since all lengths parallel to the motion in the moving frame are contracted by the same factor, R, we may think of this as space itself in the moving frame being contracted. Even the distance between two points will be contracted; we do not need an actual physical object to contract. It is possible to get confused when we think of the contraction this way. For example, we may think that light might have its speed of travel measured as faster than c because the meter sticks in the moving frame have contracted. But there is no problem; the clocks at rest in the moving frame all run slow by the same factor that the lengths were contracted, so the moving observers will calculate the same speed for light that we will. To avoid confusion, it helps to simply remember Theorem 1, that light travels at c relative to *any observer*.

One final point about length contraction: A stick that is seen by us to be moving in the direction of its length will be shortened by the factor R. We do not need to refer to its rest frame. We can define a rest frame for the stick, but we do not need to use it in this case. How the contracted stick will actually *look* to us is one of the topics covered in Chapter 6.

1-14 Summary So Far

In addition to **Postulate 1, Postulate 2** and **Corollary 1** (Section 1-5), we now have:

1. Light moves at c relative to the observer (or experimental apparatus). (This is Theorem 1)

2. A clock moving at a speed v relative to an outside observer will be determined by that outside observer to be running slow by.

$$R = \sqrt{1 - v^2/c^2}$$

That is, in a moving frame time itself runs slow as determined by an outside observer.
3. As determined by an outside observer, all lengths parallel to motion are contracted by the same factor, R, but lengths perpendicular to the motion are unaffected (perpendicular lengths are absolute or non-relative).

We will now use the first and third of these in another gedanken experiment. This experiment will further confirm the second statement above.

1-15 The Parallel Light Clock

This section confirms (adds evidence in support of) a result that we already have from Section 1-9, The Transverse Light Clock. The result that we will further confirm is point 2 in Section 1-14 that clocks, and thus time itself, runs slow by the relativity factor, R, as determined by an outside observer. More importantly, we will also gain some valuable experience.

Figure 1-10 shows a light clock that is moving past us at a speed v. Unlike the previously considered *transverse* light clock, in this experiment the mirrors are moving parallel to the directions of the light pulse that bounces between the mirrors. As before, the mirrors are separated by a distance L, as measured by an observer moving with the mirrors. Also, as with the transverse light clock, the observer moving with the mirrors will calculate that the time between ticks (tick-tock-tick, *not just* tick-tock) is $t' = 2L/c$ (time = distance / rate). In her (or anyone's) frame, everything at rest always seems normal. Since she is at rest relative to the light clock, there is no internal direction of motion (except for the bouncing light pulse) to be parallel to. (In this experiment we will say that "ticks" occur at the left mirror, and that "tocks" occur at the right mirror.)

For the remainder of this discussion, we will be describing the events as determined in our frame (in which the mirrors will be moving to right at a speed v). As *we* determine it, the distance between the mirrors is contracted to:

$$L\sqrt{1 - v^2/c^2} = L\,R$$

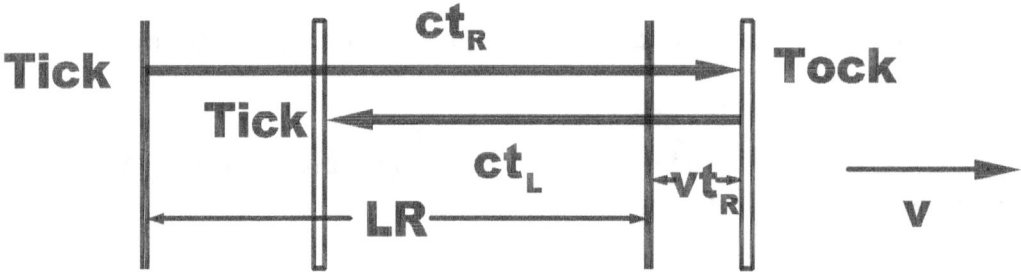

Figure 1-10. The Moving Parallel Light Clock, c shown in *our* frame where the mirrors are moving to the right at a speed v. The heavy dark vertical lines are the positions of the mirrors the first time (during our analysis) that the light pulse reflects off the left mirror. The tall hollow box at the far right is the position of the right mirror when the light pulse reflects off it. The tall hollow box at the left is the position of the left mirror when the light pulse *next* reflects off the left mirror.

Let t_R be the time it takes for a light pulse to move to the right between the mirrors. At this point we do not know how long this time is, so we give it a name, or symbol, so that we can work with it. In the time, t_R, the light travels to the right a distance ct_R (distance = rate times time). From Figure 1-10 we can see that this distance, ct_R, is also equal to the (contracted) separation between the mirrors, LR, *plus* the distance, vt_R, that the right mirror moves to the right during this time interval, t_R. It is important that the reader makes the effort to understand this bit of logic. Similar steps will be taken later. Thus, we can write: $ct_R = LR + vt_R$

Solving for t_R (See Appendix (A2-3): $t_R = \dfrac{LR}{c-v}$ (tick-tock time)

Now let t_L be the time it takes for the light pulse to move to the left between the mirrors. In this time the light travels a distance $ct_L = LR - vt_L$.
Similarly, this distance is equal to the contracted separation between the mirrors *minus* the distance, vt_L, that the left mirror moves to the right (toward the approaching light pulse) during this time interval. That is (Again, make sure that you follow this logic which gives us the next equation.):

Solving for t_L (See Appendix A2-3): $t_L = \dfrac{LR}{c+v}$ (tock-tick time):

The total time between ticks, t, (tick-tock-tick as determined by *us*, in *our* frame) is then (See Appendix A7-1 for the manipulations to obtain the final expression in Equation 1-3). Remembering

$$t = t_R + t_L = \frac{LR}{c-v} + \frac{LR}{c+v} = \frac{2L}{cR} \qquad \text{1-3}$$

that the time between ticks as determined by the observer moving with the mirrors is $t' = 2L/c$, we get from Equation 1-3 that $t = t'/R$ or $t' = t/R$:

(For this last form we multiplied both sides of the previous equation by R and turned the equation around.) **Thus, *in our frame*, we determine that the moving parallel light clock runs slow by the same** (now familiar) **factor, R, as does the transverse light clock.** Notice that length contraction in the direction of the motion (LR) was necessary to arrive at this conclusion. Remember,

$R = \sqrt{(1-v^2/c^2)}$.

We could have logically arrived at this conclusion directly from Postulate 1. If simply rotating a light clock caused it to run at a different rate (to us), then we would have detected that the clock was in absolute uniform motion. We went through the analysis of this gedanken experiment because this will be very helpful experience for the reader in upcoming gedanken experiments.

The tick-tock time, t_R, for the parallel light clock is longer than the tock-tick time, t_L, as determined by us. During t_R the target mirror (the right mirror) is moving away from the light, while during t_L the target mirror (the left mirror) is moving toward the light. We can get the same result from the math: These two times have the same numerators, but t_R has the smaller denominator. Thus, this clock would run "tick . . . .tock tick . . . .tock tick . . . .tock tick.. ." as determined by us. This is similar to a pendulum clock that is leaning to the side. However, we have shown that the time between ticks or the time between tocks for the *parallel* light clock (that is, one complete cycle) will be the same as with the *transverse* light clock. As *they* would determine the situation, their parallel light clock would not be moving, so they would determine that the light clock would run smoothly and at its normal rate. If they had a light clock that was at some angle other than $0°$ or $90°$ to the motion, its action *to us* would be a combination of the transverse light clock and the parallel light clock. To us, the light would follow a sawtooth pattern. Postulate 1 assures us that the light clock will run at the same rate regardless of its orientation. If it did not, then we could detect absolute uniform motion by simply noting that two light clocks in the same reference frame that are at some angle to each other run at different rates.

We have already begun to see that motion faster than the speed of light is not allowed in nature. However, do not be concerned when we write $c + v$, as in the tock-tick time because we are not claiming that anything is moving at this speed. We will also see this in later chapters.

Our discussion above brings up this question: Suppose that we draw a straight line segment at some angle on a playing card. Will the angle of the line be the same in a reference frame moving parallel to one edge of the card? Explain. (Hint: Think of the angled line as the hypotenuse of a right triangle. We then have enough results to answer the question. The answer is in Appendix A4-3)

1-16 An Actual Confirming Observation
We have developed several interesting results from our postulates. Is there any evidence that can be cited to support this theory? There is! We will describe the ongoing process that cosmic rays strike Earth's upper atmosphere and produce other particles called "muons" which rain down on Earth's surface. Without the effects from relativity, these muons could not reach Earth's surface. They wouldn't live long enough to make the trip down through the atmosphere. Some other experiments which are more direct, such as flying atomic clocks in orbit also confirm the results from relativity. The giant particle accelerators can produce very fast atomic particles which clearly show relativistic effects, but these effects are mostly about physics things, so discussions of these experiments will have to wait until later. The cosmic rays probably produce the most interesting experiments that we can fully describe at this point in our development, so here we go.

Cosmic rays are mostly protons (the positively charged particles in the nuclei of atoms) that have been accelerated to speeds very near the speed of light by supernova explosions and other processes that are less well understood. For such tiny particles, they possess very great kinetic energies (energy of motion). Almost all of them that approach Earth are blocked by our upper atmosphere and almost none reach sea level, but some do reach mountain tops. When they strike the nuclei of atoms in the upper atmosphere, nuclear reactions occur which primarily produce particles called muons (similar to electrons, but much more massive) which rain down at great speeds mostly reaching the ground. Muons are unstable particles. In their rest frame, they have a lifetime of only about 2.2 microseconds (2.2 millionths of a second). This is as measured by an observer at rest relative to a muon before it decays (self destructs) into other particles. (See Appendix A1-5 for the definitions of prefixes such as "micro.") 2.2 microseconds is a kind of average time (called a half-life) for muons to decay away. Some muons will decay in less time, others will take longer to decay. Even if the muons were traveling at c, they could only travel (without relativistic effects) about 2.2 microseconds times c (300 million meters per second times 2.2 microseconds) which gives 660 meters, before decaying away. Now Earth's atmosphere is about 100 kilometers thick, yet a very great number of these muons do reach the ground, producing one of the major sources of the background radiation that we live with our whole lives. How do we, at rest on the ground, explain the muon's ability to travel so far? With relativity it is easy: As we determine it, the muons are moving at nearly the speed of light, so in our frame, their life time passes much more slowly than in ours. That 2.2 microseconds gets stretched out greatly, making it possible for the muons to have time to reach the ground before decaying away. To produce such a large effect the muons must be moving about 0.9999998 c. (To achieve the slowing of time by the factor of 100 on the Questorian spaceship in the opening paragraph of this chapter, it would have to be moving at 0.99995c.) See Table 1-1. This extreme speed is only required for muons that are produced about 100 kilometers up. This height is usually defined as the edge of space. More muons will be produced lower where the atmosphere is denser. These muons will be able to reach the surface traveling at a lesser, but still very great speed.

How would an observer that is moving along with a muon explain the muon's ability to reach the ground? To this observer the muon is at rest, so it lives only 2.2 microseconds. But by Corollary 1, this observer sees Earth rushing toward him at 0.9999998 c. Imagine something as large as Earth hurtling toward you at this speed. Perhaps the muons die of fright. Seriously, as seen in the muon's rest frame, Earth is moving so rapidly that all its longitudinal (parallel to the motion) lengths will be contracted greatly. As the muons determine the situation, they live their normal lifetime, but Earth's atmospheric thickness on the hemisphere moving toward the muon is greatly contracted (to about 660 meters), allowing the *ground to hit them* before they decay away. (The atmospheric thickness on the opposite side of Earth and the solid ball of Earth are also contracted, but this does not affect the muons in question.)

Relativistic effects really are observed. An important point is that while we on the ground and an observer moving with a muon *disagree* about *how* the muons and the ground happen to get together, the only fact that must be explained is that they do, in fact, get together. Remember, only relative motion is real. There is no *absolute* uniform motion. (We have said "absolute *uniform* motion" many times now, so remember that this is what we mean when we simply say "absolute motion." Remember, "uniform" means no change in speed or direction of motion.)

An observer moving at some intermediate speed between the two observers that we just considered would find that the muons move very rapidly, perhaps 0.9999c, past her heading toward Earth, and also she would determine Earth to be rushing toward her at nearly the speed of light. (In our study we have yet to develop the means to calculate just how fast, but we will later.) This observer would find a combination of slower time for the muons and some reduction in the depth of Earth's atmosphere, and would also determine that muons and the ground do get together.

To close this section it should be emphasized that the muons penetrate Earth's atmosphere fairly easily; it is their limited lifetime that might prevent them from reaching the ground. An example where penetrating a substance imposes a limit is the case of a bullet moving at 0.9 c being shot at a person wearing body armor. In any frame of reference, the bullet will not decay away. Consider the situation from the rest frame of the armored person: Let us assume that the bullet will be stopped by the armor, so the wearer will live. But now consider the same situation from the rest frame of the bullet. In this frame the person wearing the armor will be moving at 0.9 c toward the stationary bullet. This means that the thickness of the (front and back) body armor will be contracted (thinner). Does this mean the wearer might be killed in this frame? This would certainly be a serious contradiction. The resolution of this paradox is that when the armor thins because of its motion, it still has the same number of atoms in it, so the density (weight per volume) of the armor material will increase. This makes for tougher armor, so the wearer will also live in this frame of reference. We will also see later in our study that neglecting to consider all of the relativistic effects, or in this case, failure to consider some resulting physical effect, can lead to apparent contradictions.

1-17 Mr. Tompkins

George Gamow wrote a series of short stories (from 1939-67) about a student named Mr. Tompkins who would fall asleep in physics class. (Obviously these stories were fiction.) Mr. Tompkins would then dream that he was in a land where something that the instructor was talking about was changed in some extreme way. When the instructor was talking about relativity, Mr. Tompkins dreamed that he was in a land where c was a very small value so that relativistic effects were obvious in everyday occurrences. For example, one could approach the speed of light on a bicycle. In the bicycle rest frame, peddling harder did not cause the rate that the scenery passed by Mr. Tompkins to speed up much because he was already moving nearly at c, but he got to his destination quite a lot quicker anyway because the distance to the destination, in Mr. Tompkins' frame, was even more contracted than at the lower bicycle speed. People beside the road explained this phenomenon as Mr. Tompkins' time was running slower so it seemed to him that he got there a lot sooner when he peddled harder and moved a little faster even though the distance was the same (to them). This sounds very much like the cosmic ray muons in Section 1-16. (It is!)

Would relativity be any different if c were a different value, say 30 miles per hour? The factor that enters into our considerations *because of relativity* is always the fraction, v/c. That is, the fraction of the maximum allowed speed. The modifications to "non-relativistic" equations that we develop will never depend on the actual value of c, only this fraction. But if we ask some question such as, "By what fraction are longitudinal lengths contracted at 100 million miles per hour?", then the value of c *does* enter because in this case the value of v/c does depend on the value of c. Notice that this question is not complete. It should have been something like, "As determined by observers in Frame A, by what fraction are parallel (or "longitudinal") lengths contracted in Frame B which is moving at 100 million miles per hour relative to Frame A?" Remember the admonition in Section 1-9 about always specifying the reference frame of the question, the argument, the figure, the calculation, or the answer.

We now have the basics of relativity. We will next proceed to apply what we have found to this point and arrive at additional, and perhaps even more interesting and unexpected truths about our universe. The simple Postulates 1 and 2 apparently have a lot to say.

Chapter 2 − Simultaneity, Clock Synchronization, and Velocities

2-1 The Relativity of Simultaneity

We have already seen that moving clocks run slow and that moving objects are shortened in their direction of motion. Since time is that which is measured by clocks, then time itself slows by the factor, R, in a moving frame. We may be getting used to these effects by now, or we may have heard of them before. In any case these do not seem as outrageous as the topic we will now consider. We will find that two events that are simultaneous in one frame are not necessarily simultaneous in a frame that is moving relative to it. To make matters even stranger, under certain circumstances the order of occurrence of two events will not be the same to two different observers. We are not talking about *seeing* two events at different times because they are at different distances from us and thus the signals from the events take different times to reach us; we assume that every observer corrects for the signal travel time before reporting his results. For example, if we see a fender-bender in the observatory parking lot at 11:00 p.m., and then see a supernova explosion at midnight in a star cluster that is10,000 light years away from Earth, we would *not* conclude that the accident occurred before the supernova. We would conclude that the supernova occurred 10,000 years minus one hour before the wreck. [A "light year" is the *distance* that light travels in one calendar year (365.2422 days.). It is *not* a long *time* interval, even though it is often used this way, nor is it a year when things were easy, nor a year with a third fewer days.]

Important point: The inclusion of corrections for the signal travel times constitutes the distinction between the words "determine" or "find" and the word "see." Our eyes, or a camera, simply work with the light that is arriving at the moment. This is "seeing." When we consider most of our gedanken experiments, such as the one shown in Figure 2-1, we are taking a "god's-eye" viewpoint where everything in the drawing is essentially *at the same distance* from us so that no correction for signal travel times will be needed because they all have the same delay. This frame of reference will usually be described with some possessive pronoun such as "our," or "her," or "as seen by us" because whoever is at rest in this frame is *backed far off* looking at the experiment (or page)

Consider the situation in Figure 2-1. Two identical spaceships pass by *us* in opposite directions with equal speeds (until further notice, all of this description will be as determined in *our* rest frame). Each ship has an astronaut [Ava (A) or Bob (B)]. They have used measuring tapes to determine that they are equidistant from the two **round** windows in their own ship's rest frame, located fore and aft. Just as the windows line up, flashes occur at the points labeled 1 and 2. These flashes

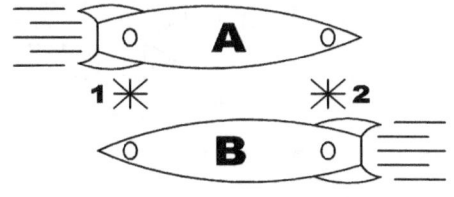

Figure 2-1. In *our* frame: Two identical ships passing with equal speed. Flashes go off when the windows align.

might be caused by matches attached to the ship's window frames that strike each other as they pass. To *us*, in our frame, the two flashes will be simultaneous. We will see the flashes simultaneously, and because we are equidistant from the two flashes, we will also *determine* that the flashes are simultaneous.

We will now discuss what Ava and Bob see and determine as to the order of Flashes 1 and 2; we actually know enough relativity now to be able to do this. The first discussion will be the more difficult to follow, and probably the less convincing. The second will be done with additional pictures. We could call a picture a "kiloword"; we will see if the additional pictures really do save a thousand words. First discussion: (In *our* rest frame) Refer to Figure 2-1. Ava is moving toward the light pulse from Flash 2, but away from the light pulse from Flash 1. Light moves at *c* relative to the observer (us), so we determine that Ava will see the light pulse from Flash 2 before the light pulse from Flash 1. Similarly, we determine that Bob will see the light pulse from Flash 1 before the light pulse from Flash 2. Meanwhile, by using their measuring tapes, Ava and Bob have determined that they are equidistant from the two flashes (windows). Thus, they need not make any corrections for the signal travel times if they are only interested in the order of occurrence of the two events. Ava and Bob will not only *see* the flashes in opposite order, they will *determine* that the flashes occurred in opposite order because the windows (flashes) are equidistant to both Ava and Bob. Thus, to Ava, Flash 2 occurred before Flash 1, while Bob will find the opposite order. Flashes 1 and 2 are simultaneous to us, but not to the two astronauts who are moving relative to us. This type of situation might enter into legality in an Old West gunfight, at least the way that movies depict such things. Who drew his gun first might be seen, and determined, as opposite to two witnesses strolling down the sidewalk in opposite directions. Yet both can be correct (in their own frames).

Second discussion: Consider Figure 2-2. The situation in Figure 2-1 as determined in Ava's rest frame is shown in the upper half of Figure 2-2. To Ava, her ship is not contracted as it is to us,

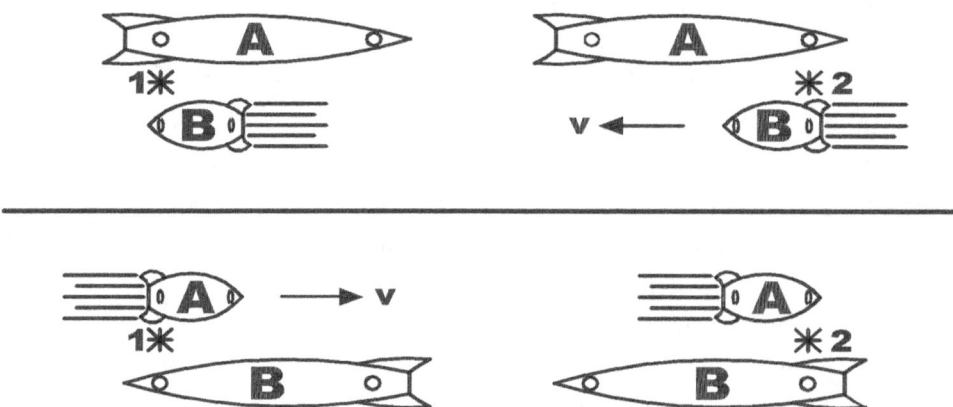

Figure 2-2. The passing spaceships: (above) in the rest frame of Ava; (below) in the rest frame of Bob

but to her, ship B will be moving to the left at a speed even faster than it moves relative to us. We have yet to show this to be true, but we will later in this chapter, and it turns out that it is *not* twice the speed. Because of its greater speed to Ava than to us, the length of ship B will be even more contracted to Ava than it is to us. From the upper half of Figure 2-2 Ava will clearly determine that Flash 2 occurs before Flash 1. This is because the right pair of windows and matches in Figure 2-2 line up before the left pair. So from Ava's frame, Flash 2 really does occur before Flash 1. Since Ava is equidistant from the two flashes, she will even see Flash 2 before Flash 1. (The matches attached to the window frames are aligned so that the other two apparently-possible flashes do not occur, only the shown flashes occur.) A similar argument will show that Bob will obtain the opposite order (refer to the lower half of Figure 2-2). We may have saved a few words because of the additional pictures, but not a thousand. However, more importantly, wasn't the second discussion the more convincing? Notice that length contraction of the moving ship is shown in Figure 2-2 and that this was essential for this argument. Also, the round portholes (in the rest frame of a ship) are ellipses (ovals) in a frame that sees the ship as moving. The labels, A and B, are drawn in the figure, rather than being painted on the ships, which is the reason for their lack of contraction (as well as making for easier readability in the figure).

The second discussion method has demonstrated some of the power of considering the same situation from more than one frame of reference. In some of what we will be doing, the use of more than one reference frame will be necessary, not merely helpful, to arrive at our conclusions. In Section 1-1 it was mentioned that relativity involved viewing the same process from more than one frame of reference, and that this did not sound too interesting. As promised then, very surprising truths will emerge from such considerations. We have only barely begun.

2-2 Clock Synchronization

Before we further investigate these, and additional effects, it will be helpful to consider how we can synchronize two clocks that are at rest within the same reference frame. Figure 2-3 shows the "flash in the middle" method of synchronizing a pair of clocks: The clocks are at rest in Ava's rest frame and paused at the same setting (usually zero – this is what is shown in the figures). A stick, not shown, holds a flashbulb at the midpoint between the clocks. The bulb fires. When the flash arrives at a clock, it starts running. Clearly, this method will synchronize the clocks. Appendix A8-3 describes another method of clock synchronization called (here) the "WWV method." We will need and describe the WWV method later.

Note: Beginning at this point, much of the logic

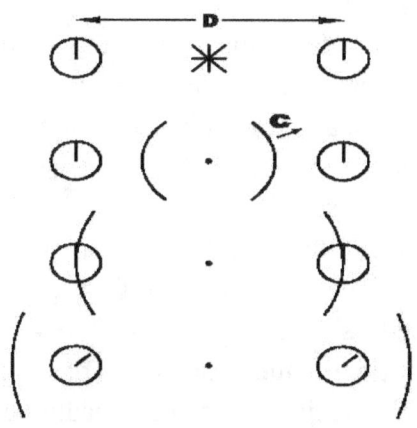

Figure 2-3. Synchronizing two clocks by the "flash in the middle" method by an observer, Ava, who is at rest relative to the clocks. The black dot is the burned out flash bulb.

will become more quantitative (numbers and symbols) than it has been. It is only possible to do the quantitative work if we use the logic of numbers, that is to say, math. Do not be distressed. The reader needs only to follow the steps that are written and described. There are some simple math lessons in the appendices. There the manipulations required for the simplifications are also shown and explained, step-by-step. The math lessons in the appendices are adequate, but quite brief. A good source that is more detailed is the early portion of Bobrow, listed in "Suggested Readings." The rewards for following the logic are great. If one also wants to follow the manipulations involved in some of the steps, or even works them out herself, then the rewards are even greater. A companion note to this one follows Equation 2-5 in Section 2-3. Remember the quotation from Einstein that is in the preface, "Everything should be as simple as possible, but not simpler." Any relativity book that has little math in it is simply "describing" relativity, rather than *discovering* or *revealing* relativity. We are actually *deriving* relativity from the postulates. The math will be kept as simple as possible.

We saw one method of synchronizing two clocks in Figure 2-3, but how will an observer, Bob, who is moving relative to the clocks, describe this process? Figure 2-4 shows the important events as determined by Bob, who is moving to the *left* at a speed, v, relative to the rest frame of the clocks. Notice that even though the clocks show the effect of length contraction, the light still forms a hollow sphere, not an oval, of photons because light moves at c in all directions in any reference frame. This is Theorem 1.

For the remainder of this section we will be considering this process of clock synchronization in Bob's rest frame. To Bob the clocks are moving to the *right* with a speed, v, and the separation between the clocks is contracted to DR. (R gets smaller as v gets larger.) The time when the flash goes off will be defined as zero seconds in Bob's time. Notice

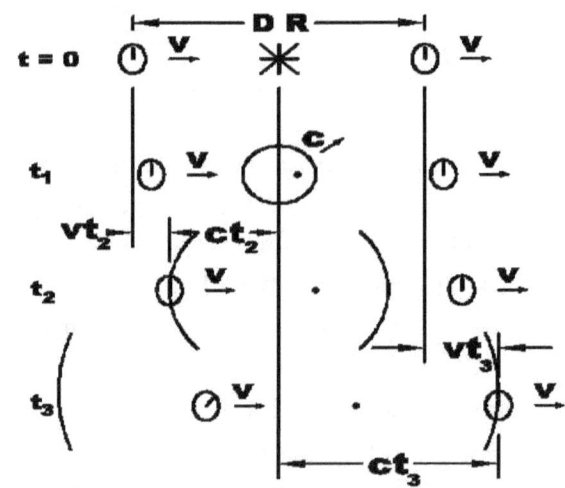

Figure 2-4. Synchronizing two clocks by the "flash in the middle" method by an observer, Bob, who is moving to the left at speed v relative to the clocks. The burned-out flash bulb (black dot) remains centered between the clocks.

that at time t_2 the light signal (flash) has reached the "following clock" (on the left in Figure 2-4), so it starts running before the "leading clock" does. Thus, even though the clocks are synchronized to Ava, the clocks will *not* be synchronized to Bob who is moving relative to the clocks. In Bob's frame, the following clock, which starts running earlier, will read ahead of the leading clock (on the right). In any observer's frame the light still moves at c and spreads out in a spherical wavefront centered at the point of the flash. This comes from Theorem 1 and will be further discussed in Chapter 3.

How much ahead will the following clock be to Bob? Analysis very similar to what we have already seen in Section 1-15 will provide the answer: At t_2 light reaches the following clock and it starts. As yet, we do not know what t_2 is, so we give it this label so we can work with it. During this time, t_2, the light flash will have moved a distance ct_2 toward the following clock, and the

$$ct_2 + vt_2 = \frac{DR}{2} \implies t_2 = \frac{DR}{2} \frac{1}{c+v} \qquad \text{(See Appendix A7-2)}$$

following clock will have moved a distance vt_2 to the right (toward the light pulse). These two distances add to half the contracted separation between the clocks. (See Figure 2-4.) That is:

The following clock starts at time t_2, but the light pulse has not yet reached the leading clock so it is still paused at zero. Let t_3 be the time when the light pulse finally reaches the leading clock. During this time (from zero) the light pulse has moved to the right a distance ct_3. The leading clock has moved to the right a distance vt_3. The distance the leading clock moves to the right plus half the contracted separation between the clocks equals the distance the light travels in reaching the leading (right) clock. Make sure that you see Figure 2-4 on this. That is:

$$vt_3 + \frac{DR}{2} = ct_3 \implies t_3 = \frac{DR}{2} \frac{1}{c-v} \qquad (\text{See Appendix A7-2})$$

The difference between these two times is:

$$t_3 - t_2 = \frac{DR}{2} \left(\frac{1}{c-v} - \frac{1}{c+v} \right) = \frac{Dv}{c^2 R}$$

However, during this time interval (still in Bob's frame), the following clock has been running slow by the factor R, so the following clock will be ahead of the leading clock by only

$$\frac{Dv}{c^2 R} \cdot R = \frac{Dv}{c^2} \qquad (2\text{-}1)$$

(See Appendix A7-2)

We will call this final fraction "**Result 2-1**". It is important to note that D is the rest (uncontracted) spacing between the two clocks as measured in the rest frame of the clocks. We will use Result 2-1 many times in later gedanken experiments.

33

Exercise 2-1: Devise a gedanken experiment (i.e., draw a figure similar to Figure 2-3, but with vertical motion instead) to show that when using the "flash in the middle" method of clock synchronization that clocks separated transversely (at right angles) to the motion of another observer will be determined to be synchronized in both the rest frame of the clocks and in the moving frame. No math is needed. The solution is shown in Appendix A8-1.

Exercise 2-2: Now use the result of Exercise 2-1 to expand the application of Result 2-1 to the case where the clocks are separated at an angle to the line of motion of the moving observer. See Figure 2-5. Hint. Use Result 2-1, a properly placed third clock (two different places will work), and the result from Exercise 2-1 to find the answer. (No math is required.) What will D be here? A solution is shown in Appendix A8-2. Make sure to at least look at the solution because we will need this result later.

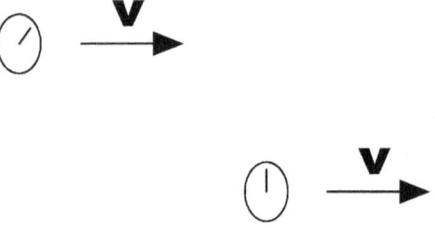

Figure 2-5. The clocks from Exercise 2-2 as determined by an observer moving to the left at a speed v.

Notice that in Figure 2-4 the light sphere stays *centered where the flash occurred* in Bob's frame and does *not* stay centered on the burned out flashbulb which to Bob is moving to the right. This result is from Theorem 1, and it will be discussed more completely in Section 3-2. The light sphere stays centered at the position of the flash in Ava's frame too, and also in every other frame, regardless of the motion of those frames, because light moves at c relative to all observers.

An alternative method of synchronizing clocks, which we will call the "WWV method," is described in Appendix A8-3. The reader should refer to it at this point or during Section 2-4.

2-3 Velocities and 1-Dimensional Velocity Combinations

You may have heard some people say that relativity cannot be correct in claiming that nothing can go faster than the speed of light. Their argument might describe a gedanken experiment where a spaceship is moving to the right at 90% of the speed of light (0.9 c), and another spaceship is moving to the left at 90% of the speed of light. The claim is then made that as determined by one of the ships the other ship is moving at 1.8 times the speed of light. Their assumption is that approaching velocities simply add at all speeds as they seem to at the very low speeds that we encounter daily. The reader might not be too surprised to learn that this is not a real problem for relativity, with all its time and length effects and disagreements about clock synchronization.

However, not realizing that speeds of approach may not simply add delayed the discovery of relativity for several years.To investigate this situation, consider Figure 2-6. As *we observe it*, a spaceship is moving to the right at a speed V, and some particle is moving to the left at a speed v.

The question is, "What speed will an observer on the spaceship determine for the particle?" We must be very explicit about how we will make this speed measurement. Such measurements usually require several steps, and we must apply the appropriate relativity effects at each step.

We begin: The spaceship is equipped with clocks located in the nose and tail which were synchronized in the rest frame of the ship to record the times of passage of the particle by the nose and tail. Ava is on the ship and has carefully used her measuring tape to measure the distance, D, between the two clocks in her rest frame. She has radioed this clock-separation information to us in digital form. We will use what we now know about clocks running slow, clock synchronization, and length contraction to predict what measurements Ava will obtain, then calculate the velocity that she

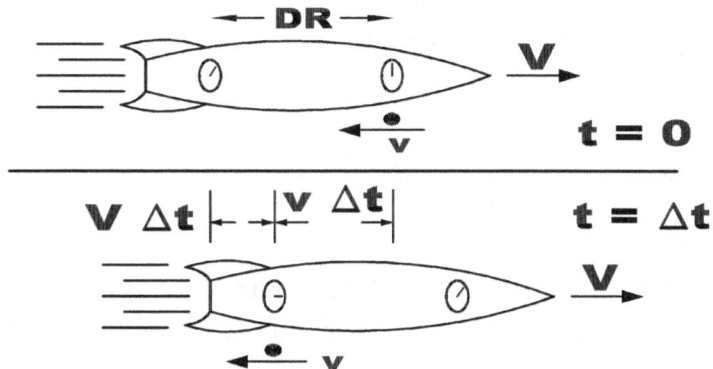

Figure 2-6. As determined in *our* frame, the ship is moving to the right with a speed V, and the particle is moving to the left with a speed v. The upper half of the figure is at the moment that the particle passes the nose clock, while the lower half is at the moment that the particle passes the tail clock.

will calculate from these measurements. We know enough to do this. As we determine it, the distance between the clocks is contracted by the usual factor, R.

Here R must be calculated using the speed of the spaceship, V, not the speed of the particle, v. In our frame, the ship and all of its contents are moving to the right at V.

Let the clock in the nose read zero when the particle passes the nose clock. The clock in the tail will then read DV/c^2 (Result 2-1); **the following clock, synchronized in the rest frame of the two clocks, always reads ahead of the leading clock by this amount.** Let Δt be the time interval to us, in our frame, for the particle to go from the nose clock to the tail clock. During this time interval, Δt, the particle moves to the left a distance $v\Delta t$, and the tail clock moves with the ship to the right a distance $V\Delta t$. When added, these two distances equal the contracted separation, DR, between the two clocks (See Figure 2-6). That is:

35

$$VΔt + vΔt = DR$$

Solving the previous equation for $Δt$:

$$Δt = \frac{DR}{V + v} \qquad \text{(2-2) (See Appendix 9)}$$

During this time the clocks on the ship have been running slow by the factor R (to us in our frame), so the tail clock will have *advanced* by only:

$$Δt \cdot R = \frac{DR}{V + v} \cdot R = \frac{DR^2}{V + v} \qquad \text{(tail clock advancement to \textit{us})}$$

Let t' be the reading of the tail clock when the particle reaches it. At this moment the tail clock will read (the amount it reads ahead of the nose clock in our frame plus its advancement to us):

$$t' = \frac{DV}{c^2} + \frac{DR^2}{V + v} \qquad \text{(2-3)}$$

To Ava, the particle moves a distance D, and the time difference between her clocks is this t', so she will calculate the speed of the particle to be $v' = D/t'$, that is:

$$v' = \frac{D}{\dfrac{DV}{c^2} + \dfrac{DR^2}{V + v}} \qquad \text{(2-4)}$$

Remember that in this gedanken experiment R is calculated using (upper case) V, the speed of the ship as determined by us. So here, $R^2 = \left(1 - V^2/c^2\right)$. In Equation 2-4, D divides out of the numerator and the denominator (see Appendix A1-2), showing that the separation between the clocks is unimportant, but it must not be zero.

Simplifying Equation 2-4 for the speed of the particle in Ava's (the ship's) frame, yields (see Appendix 9):

$$v' = \frac{V + v}{1 + \dfrac{Vv}{c^2}} \qquad (2\text{-}5)$$

Except for the second term in the denominator, Equation 2-5 is precisely what we would expect when combining approaching velocities without relativity, that is, simply adding them. For low speeds (say less than 20,000 miles per second) this second term in the denominator will be very small compared to 1. However, as both v and V approach c, then this second term approaches the value of 1, so the complete denominator approaches the value of 2. It is this second term in the denominator that prevents the speed combination from ever giving a calculated speed greater than c.

There have been so many logical steps beginning with the postulates to this point that trying to follow the logical arguments all this way would have been extremely difficult without the use of math to keep up with the logic. Also, using only words it would have been essentially impossible to have arrived at a quantitative result. Most readers will probably want to see the derivation of the most famous equation in the world. To do this, Equation 2-5, or rather its slightly adjusted form, Equation 2-6, among others, will be essential in this endeavor.

Equations 2-4 and 2-5 were found by using speed = distance / time. As determined in *our* frame, three effects from relativity entered the derivation of Equation 2-5. These are (in our frame): 1) The distance between the clocks is contracted. This will *decrease* the *calculated* speed because the distance over which the particle is timed is reduced. This reduces the numerator in speed = distance / time. 2) The clocks run slow during the passage of the particle from nose to tail; this *decreases* the *measurement* of the time of passage of the particle between the two clocks, and thus *increases* the calculated speed because the measured time appears in the denominator of speed = distance / time, and 3) the two clocks are not synchronized in our frame; this *decreases* the calculated speed because the difference between the clock readings (tail clock minus the nose clock) is *increased* since the tail clock (the following clock) always reads ahead of the nose clock. The particle speed is found by the distance between the clocks divided by the time difference between the two clocks. The appearance of the second term in the denominator of Equation 2-5 is the total effect that relativity has on this calculation. This term *decreases* the calculated speed. We might have guessed this from the three effects mentioned above; two effects decreased the calculated speed, but only one effect caused an increase. However, without using math to do the logic we could not know which of these effects are large and which are small. Thus, claiming that the two-to-one advantage means that the overall effect is to decrease the calculated speed is illogical. Math really is needed to help with the logic.

Corollary 1 requires that the particle determine the same speed for the approaching spaceship as the spaceship does for the particle. We can check Equation 2-5 for this by simply exchanging V and v. V would then be the particle speed, and v would be the speed of the ship, both in our frame. From Equation (2-5), notice that this does not change the calculated value for v' (now the ship's speed in the particle's rest frame). The order of adding or multiplying numbers makes no difference in the calculation. It is reassuring to check that our new result is consistent with an earlier result (Corollary 1).

Table 2-1 shows some examples of velocity-combining using this relativistic velocity combining equation (Eq. 2-5):

Table 2-1. Combining Selected Velocities Using Equation 2-5		
V (ship speed, to us)	v (particle speed, to us)	v' (particle speed, to ship, from Eqn. 2-5)
0	0	0 (like without relativity)
0	v (any value, $+$ or $-$)	v (like without relativity)
$0.001c$	$0.001c$	$0.001999998c$ (very close to adding)
$c/2$	$c/2$	$4c/5$
$0.9c$	$0.9c$	$0.994475c$ (not greater than c at all)
V (any value, $+$ or $-$)	c	c
c	c	c

Verify the first two and the last two rows to convince yourself of these results. Notice that the *value* of c is not needed to calculate this table. The next to last row shows that if the particle is a photon ($v = c$), then the observer on the ship will determine the same speed that we do. This is true regardless of the speed of the ship relative to us, V. The last row shows that even if the ship could move at c to the right, the speed of a photon moving to the left would still be determined by the observer on the ship to be the characteristic speed of a photon, namely, c. These last two rows show that we have not made a mistake that would cause a conflict with Postulate 2 or Theorem 1. A further example: If both V and v are each one million miles per hour, then v' will be 1,999,996 miles per hour. This is only four miles per hour less than what we would have thought before we learned some relativity.

The fifth row in Table 2-1 is the situation described at the beginning of this section. Notice that the $0.9\,c$ speeds of the spaceship and particle do not give a velocity of approach that is greater than the speed of light. Velocities do not simply add! However, they do come very close to simply adding when the velocities are small compared to c.

It was both more interesting and easier to do both the logic and the interpretation of the resulting formula by considering this "head on" case rather than some overtaking case where both the ship and the particle are moving in the same direction but at different speeds. Nevertheless, our velocity combining (not adding) equation will be more easily useful in later considerations if both the ship and the particle speeds are each measured as positive in the same direction. This adjustment is most simply done by considering all velocities in Figure 2-6 as *positive to the left*. Since the *particle* velocities were already assumed to be to the left, both in *our* frame, v, and in the ship's frame, v', no modification is needed with these. However, we assumed in Figure 2-6 that the ship's velocity in *our* frame was positive to the right, so we must simply replace V with $-V$. There is nothing significant about considering positive to be to the *left*. If Figure 2-6 had been drawn reversed right-to-left, then we would have chosen positive-to-the-right. In either case we obtain for our finished expression:

$$v' = \frac{v - V}{1 - \frac{Vv}{c^2}} \qquad (2\text{-}6)$$

where: v is the particle velocity in *our* frame, V is the ship velocity in *our* frame, and v' is the particle velocity in the *ship's* frame. These velocities can be considered positive (or negative) in any direction that we choose, but we must be consistent to express all three velocities as positive in the same direction. Equation 2-6 is the (1-Dimensional) velocity-combining equation that we will need later. Important: We used a rocket ship in finding our 1-D velocity-combining equation. Realize that the ship's frame could simply be any other reference frame that moves along the same line as the particle. Equation 2-6 is for a 1-D case.

(This paragraph is a little messy and is optional.) Earlier we applied Corollary 1 to Equation 2-5 (interchanging v and V) as a validity test. It was easier at that point than to apply Corollary 1 to Equation 2-6 because we were previously considering velocities of approach to be positive, rather than positive in a certain direction. A velocity of approach would be the same velocity for both ship and particle. Where velocities in one direction are considered as positive, as in Equation 2-6, then in addition to interchanging v and V we must also reverse the sign on v because if the ship (V) and the particle (v) are approaching each other and the particle is in front of the ship, then the ship will determine a negative velocity (backward) for the particle, but the particle will determine a positive velocity (forward) for the ship. Simply put, when we consider all velocities as positive in the same direction, then v is the velocity of the particle in the ship's reference frame, while v is the velocity of the ship in the rest frame of the particle. Now you know why we did this test using Equation 2-5. Equation 2-6 does pass the Corollary-1 test, but describing the test was more difficult than it was for Equation 2-5.

2-4 2-Dimensional and 3-Dimensional Velocity Combinations

This is an optional topic that is not essential in our development of relativity. However, if the reader wants to see an exact derivation of the most famous equation in the world, and/or is interested in a more exacting approach to relativistic momentum than is developed in Chapter 4, then the results of this section become essential. It is merely an extension of the previous section, but it does get a little complicated. If you start feeling lost, then skip the rest of this section and come back to it later after you have more experience. We will reexplore the velocity combining situation of the previous section, but this time we will consider the case where the incoming particle is not moving along the same line as the spaceship.

Consider Figure 2-7. As in Section 2-3, the analysis will be done in *our* frame. We can predict what measurements and calculations Bob will make on the ship. (It is now Bob's turn to fly

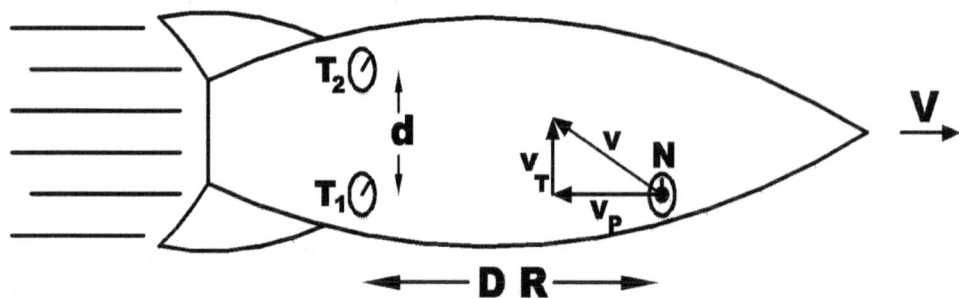

Figure 2-7. The ship and particle in *our* frame. This is the same as Fig. 2-6 except that here the particle has a vertical velocity component, v_T. All three clocks are synchronized in the rest frame of the ship. The particle is shown as the black dot as it passes by the center of the nose clock, N. The total velocity vector, *v*, is *not* pointed at the tail clock, T_2 (in our frame), but rather at the point in our frame where the particle and the moving clock, T_2, will later intersect.

the ship.) The spaceship again has a rest distance of D between the nose clock and the tail clocks measured along the axis of the ship, and the clocks are all synchronized in the rest frame of the ship. (See Appendix A8-3 on how this can be done with more than two clocks. You might find this interesting; read it after finishing this chapter.) In our frame the ship is again moving to the right with a speed of V. In this gedanken experiment the particle is not only moving to the left but is also moving upward (across the ship) as well. As in Section 2-3, horizontal velocities of approach will be considered as positive; it is simply easier to visualize the situation this way. Bob has placed tail clock T_1 straight back from the nose clock N, and placed tail clock T_2 directly across the ship from T_1 in exactly the right position so that the particle will pass through it. The angled velocity of the particle can be broken down into two components. One component is parallel to the line of motion of the ship and is called v_p "v-parallel"). The other component is transverse (perpendicular) to the line of motion of the ship, v_T, is called "v-transverse"). The particle velocity component across the ship will be considered as positive in the upward (in the figure) direction. See Appendix A4-4 for more on components.

To us (still in our frame), as in Section 2-3, if the nose clock, N, reads zero when the particle passes through it, then when the particle reaches the tail clock T_2 , both T_1 and T_2 will read (from Equation 2-3):

$$t' = \frac{DV}{c^2} + \frac{DR^2}{V + v_P} \qquad (2\text{-}7)$$

We have used v_P in place of v since this is only a new name for the same velocity as in Section 2-3. Also, synchronized clocks separated at a right angle to the motion between the frames, as are T_1 and T_2, will be synchronized to both observers (Bob and us). See Exercise 2-1 and/or Appendix A8-2.

To find v_T' , the transverse or the across-the-ship particle velocity as determined by Bob in the ship, we will use velocity = distance / time. In Section 1-12 we established the fact that distances across the line of relative motion between the two frames, that is, transverse distances, are absolute (agreed upon by both observers). Thus, the separation between the two tail clocks is d in both frames. To Bob the particle moves across (transverse) the ship a distance d in the time t . It is useful to express **d as it can be determined by distance = rate times time in *our frame*. That is:**

$$v'_T = \frac{d}{t'} \ (to \ Bob) \qquad but \qquad d = v_T \, \Delta t = v_T \frac{DR}{V + v_P} \ (to \ us \ and \ Bob)$$

We got the time Δt for the particle to both travel from nose to tail and also to cross the ship from Equation 2-2, with v again being replaced by its equivalent in this situation, v_P .

Using the above expression for d, and Equation 2-7 for t , we obtain (the mess):

$$v'_T = \frac{d}{t'} = \frac{v_T \dfrac{DR}{V + v_P}}{\dfrac{DV}{c^2} + \dfrac{DR^2}{V + v_P}} = \frac{v_T \dfrac{R}{V + v_P}}{\dfrac{V}{c^2} + \dfrac{R^2}{V + v_P}} \qquad (2\text{-}8)$$

D divides out of both the numerator and the denominator, again showing that if the separation between the nose and tail clocks is not zero, then it does not matter what its value is. Here, as in Section 2-2, the relativity factor, R, must be calculated using V, the speed of the ship in our frame. Simplification of Equation 2-8 yields (See Appendix 10, it is not as bad as it looks):

$$v'_T = \frac{v_T \sqrt{1 - V^2/c^2}}{1 + \dfrac{V v_P}{c^2}} \qquad (2\text{-}9)$$

The relativity considerations that entered into our derivation of Equation 2-9, the transverse (across the line of motion of the ship's frame) velocity transform from one frame to another are: (1) All three clocks ran slow by the relativity factor, R, during the passage of the particle. This will *decrease* the measurement of the transit time, thus increasing the calculated speed because the speed = distance / time. (2) The two tail clocks read *ahead* of the nose clock. This will *decrease* the calculated speed because the time difference between the two clocks (T_2 minus N) will be *increased*.

(3) *d*, the transverse separation of the tail clock, T $_2$, and the nose clock N is the same in both frames, so this will have no impact on the derivation. So the difference between the transverse velocity transform and the earlier longitudinal transform is that the distance of particle travel is not contracted in the transverse case.

Table 2-2 shows some representative examples of using Equation 2-5 (with *v* replaced by its equivalent, v_P, and Equation 2-9.

		Input values				Output values		
	V	v_P	v_T	v_{tot}		v'_P	v'_T	v'_{Tot}
1	0	v_1	v_2	$\sqrt{v_1{}^2 + v_2{}^2}$		v_1	v_2	$\sqrt{v_1{}^2 + v_2{}^2}$
2	$3c/5$	$4c/5$	$3c/5$	c		$35c/37$	$12c/37$	c
3	$3c/5$	$-4c/5$	$3c/5$	c		$-5c/13$	$12c/13$	c
4	$0.9\,c$	$0.2\,c$	$0.2\,c$	$0.283\,c$		$0.932\,c$	$0.074\,c$	$0.935\,c$

Table 2-2 Two dimensional velocity combining using Equations 2-5 and 2-9

Notes on Table 2-2: The magnitude of *V* must be less than *c*. The total speeds, $\sqrt{(v_p{}^2 + v_T{}^2)}$, in both frames, (i.e., both primed and unprimed) must be less than or equal to *c*. Row 1: If *V* = 0 then our frame and the frame of the rocket are the same frame, thus there will be no relativistic effects; all velocities will the same to both observers. We should perform this check to make sure that our results show no relativistic effects when the two frames are at rest relative to each other. Row 2: The particle has to be a photon because its total speed is *c* in both frames. In our frame both *v* $_P$ and *v* $_T$ are positive, thus the photon is moving upward and to the left, that is, somewhat toward the tail. In the ship's frame (the primed frame), the parallel (or longitudinal) particle speed, *v* $_p$, has greatly increased because of the head-on nature of the collision of particle and ship. Row 3: In our frame, the photon is moving partially to the right (that is, away from the ship – a negative velocity) and upward in both frames. Whatever the values of the two speed components, the total speed of a photon must be, and *is*, *c* in any frame. This agreement with Theorem 1 should give us confidence in our velocity-combining equations. Row 4: This row shows that even for equal parallel and transverse velocities in our frame, they can transform quite differently.

However, as in Section 2-2, while this gedanken experiment was easier to analyze, and the situation shown in Figure 2-7 was easier to visualize if we considered a collision instead of an overtaking case, the expression will be easier to use in later gedanken experiments if we express all our velocities parallel to the line of motion between the two frames as positive in the same direction.

This is again most easily done by considering all (horizontal in the figure) velocities in Figure 2-8 to the left as positive. As in Section 2-2, all that must be done to accomplish this is to replace V with V. (This does not alter V^2.) Also shown below is Equation 2-11, which is the parallel (1-Dimensional) velocity transformation, Equation 2-6, slightly revised by simply replacing v with v_p, its equivalent here.

Result 2-11 will be used to combine the components of velocities that are parallel to the line of the relative motion between the two reference frames (the x-components as shown in Figure A17-1). Equation 2-10 will be used to transform the *two* velocity components perpendicular to the line of relative motion between the two reference frames (the y- and z- components as shown in Figure A17-1). The spaceship that we have been using in this and the previous section is at rest in the *primed frame* of reference. Thus, we may now dispense with any mention of the spaceship, or even its pilot, Bob. Equations 2-10 and 2-11 will handle the transformation of any velocity from one reference frame to another. That is, we now have the velocity-combining equations needed for any three-dimensional situation. We will use them later in our study.

When the expression for the relativity factor, R, is inserted into Equation 2-8, then as shown in Appendix 10, Equation 2-8 temporarily becomes quite complicated. The fact that it simplifies down to Equation 2-10, with only the familiar relativity factor in the numerator and the same denominator that we have seen before in Equations 2-6 and 2-11, gives a ring of truth to what has been developed so far in our study of relativity.

There is another validity test that can be applied to Equations 2-5, 6, 9, 10, and 11: If we let the value of c go to infinity, then any relativistic equation should become the pre-relativistic form. This is actually quite simple to check: Any fraction that has c as its denominator will become zero as c becomes infinite. All five of these equations are fractions. Each of their denominators will become the value one (unity) with an infinite value for c. The relativity factor will also have the value of

$$v'_T = \frac{v_T \sqrt{1 - V^2/c^2}}{1 - \dfrac{Vv_P}{c^2}} \qquad (2\text{-}10) \qquad v'_P = \frac{v_P - V}{1 - \dfrac{Vv_P}{c^2}} \qquad (2\text{-}11)$$

unity (one). All five of these equations do simply reduce to what we would have thought to be true before we learned about relativity. For example, Equation 2-5, the head-on collision transform formula, becomes $v' = V + v$. Also, Equation 2-10 for the transverse velocity component (i.e., across the line of motion between the two frames) simply becomes $v'_T = v_T$, the same in both frames, also what we would have previously thought. So far, our theory of relativity has passed all the self-consistency tests that we have thrown at it.

Nature seems to be as simple as it can be and still function. However, with relativity, nature now seems to be a little more complicated, but also much more interesting than what was thought over 100 years ago. Physics texts on relativity develop the two velocity-combining equations by using the Lorentz transformation equations only. (The Lorentz transforms are derived from what we developed via the gedanken experiments in Appendix 17. The velocity-combining equations are *not* re-derived in the appendix.). This method is quick and sure, only a few math steps each are needed, but the reader develops no understanding of the nuts and bolts that go into the velocity-combining process. In this book, developing an understanding of what considerations enter into relativistic results is one of the prime objectives. Once this understanding develops, relativity does not seem weird at all.

We now have almost all of the relativity tools that we will need. In Chapter 3 we will apply what we have learned about relativity so far to investigate what effect relativity has on our understanding of other realities of nature. We will discover that relativity makes some changes that we would have had quite a difficult time believing if we were simply told that they were true. Since we are *deriving* all our results from the (very reasonable) postulates, we should have an easier time believing our conclusions.

Chapter 3 - Causality and Light Spheres

3-1 Relativistic Causality Defined

If two events, X and Y, occur, there is the possibility that event X somehow caused event Y to occur. That is, some type of signal or stimulus travels from X to Y. The signal might be something like sound, light, electricity, or a nerve impulse. Whatever the signal mechanism, the signal must be considered as part of the cause. It seems entirely reasonable to expect that the cause, X, occurs *before* or at the *same time* as the effect, Y. This concept is named "causality.*"* We will consider a light signal because light is one of the things that can travel at the maximum speed allowed in the universe, namely c. This will be further explained below. VERY IMPORTANT POINT: If X occurs and immediately emits a light-speed signal, then this signal must reach Y before, or at the same time, that Y occurs in order for X to *possibly* be the cause of Y. Stated oppositely: If the event, Y, occurs before a light-speed signal from X can reach it, then X *cannot* have caused Y.

Remember, *in relativity* causality means that a signal from an earlier event *could have caused* an effect to occur, not that it necessarily did so. This is slightly different from the meaning of "causal" in other areas of study where X did, or did not, cause Y. One more vital point: In deciding whether two events could be *causal*, it is necessary to consider only the fastest possible signal speed, namely c. Otherwise, we would not be properly considering the "*could* have caused" property in the meaning of "causality." An example: An ammunition dump blows up. Another nearby dump then explodes before the sound or blast wave could have reached it. But if the flash (light or radio waves) reached the second dump before it exploded then these events are causal (in the relativity sense – the only use of the word "causal" in this book).

This is similar to the "post hoc fallacy." ("*post hoc, ergo propter hoc*" in English means "after this, therefore because of this") This logical error is often made. An example is: A comet appears. The king dies. Ergo, comets kill kings. Probably not. He may have died from the gout, but the events *are causal in the relativity sense*. A light signal did travel from comet to king (he saw the comet) so he could have been frightened by the comet and died of a heart attack. It *is* illogical to conclude that something that happens later (winning a trip to Hawaii) was caused by some *particular* thing that happened earlier (wishing and blowing out the birthday candles). There is (almost certainly) plenty of time for a light-speed signal to travel from the cake to the drawing (at most about 0.07 seconds on Earth), so the events are causal in the relativity sense, but we must not conclude that the wish and puff *necessarily* caused the happy result.

One of the important issues that we need to investigate is whether or not relativity, with all of its space and time effects, can interfere with causality. In other words, if two events are causal in one frame, are they necessarily causal in all other reference frames moving relative to the first frame? And vice versa, if two events are non-causal in some frame, will they be non-causal in all other reference frames?

It seems like it would be very difficult to answer such questions since events can have all sorts of separations and motions in space and times of occurrence, and observers can have all sorts of motions and positions relative to each other. Fortunately, an expanded version of Theorem 1 that we will call the "Light Spheres Theorem" gives us a good start on the answer.

Someone looking into relativity for the first or second time probably would not have even thought to ask about the effect of relativity on causality. I'll confess that while answering this question is important, it is really a secondary goal for us. The primary reason for this chapter is to have a goal to prompt us to get some important practice in using what we have developed so far. As it turns out, in investigating the relativity-causality connection, we will need everything that we have discovered so far. We will also learn about some additional facets of relativity.

3-2 Light Spheres

By "light spheres" we mean the following: If a *flash* occurs at a point in space, the light pulse from the flash will spread out in all three dimensions forming an ever growing *hollow* spherical shell of photons whose radius is ct, where t is the time since the flash. See Figure 3-1. If another flash occurs at a later time anywhere within the first light sphere then it will form a second growing light sphere, but since their radii are both growing at exactly the same rate, c, no portion of the later (smaller) light sphere will ever catch up to any portion of the earlier (larger) light sphere. See the left

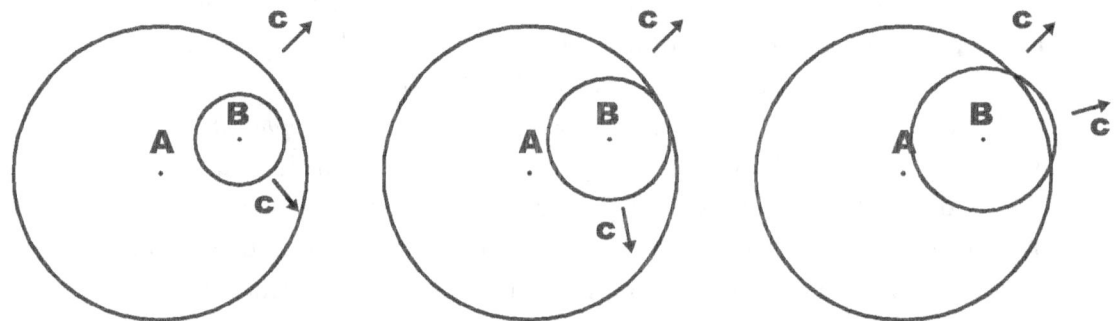

Figure 3-1. From left to right, these pairs of light spheres are: nested, barely nested, and non-nested. These circles are cross-sections of the light spheres through their centers. That is, the page has been positioned to contain both points A and B.

pair of light spheres in Figure 3-1. If one light sphere is completely within another light sphere, as is this pair, then we will refer to them as "nested."

The black circles (the closed curved black lines rather than their interiors) in the figures are where the photons are at the moment. The photons do not fill the interior of a sphere unless that light bulb stays on. Since we will be dealing with flashes, or pulses, the light spheres will be *hollow* shells of photons that are moving outward at c. The circles, or portions of circles (arcs), in all of the figures in this section are 2-D cross-sections through the *centers* of the 3-D spheres (where the flashes occurred).

This next part is a little surprising. Postulate 2 says that the speed of light is not affected by the motion of its source, so any motion of sources A or B will not affect the light spheres. The only things that affect the light spheres are the *positions* of the sources at the times when the flashes occurred and how long it has been since the flash occurred. Theorem 1 (Light moves at c in all reference frames.) tells us that the centers of the light spheres will remain fixed at the location *where the flash occurred in each reference frame*, and not fixed on the objects that produced the flashes (unless that object happens to be at rest in that particular frame). If the light spheres followed the motion of the sources to any extent, then the speed of that light in one direction would be greater than c, while in the opposite direction the speed of that light would be less than c. This would be a violation of Theorem 1.

Examine the left pair of light spheres in Figure 3-1. Realizing that both spheres are growing at the rate c, then it is clear that the flash from event A had already passed by the location of B before the flash at B occurs. This is most easily seen by considering the direction that goes from A toward B. One can see that light sphere A had already passed by B before Flash B occurred by considering the separation between the spheres in this direction. Notice that it is possible that the passage of light sphere A could have caused Flash B to occur after a short delay.

The middle pair of light spheres in Figure 3-1 shows the situation where Flash B occurred just as light sphere A reached B. The smaller sphere could even be a reflection of photons from A off B. In the direction from A toward B the spheres barely touch (are "tangent" to each other). Again, since the light spheres are both growing at the same rate, they will forever remain tangent to each other. Also again, A could have caused B, but just barely. Light spheres of this type will be called "barely nested."

The third pair of light spheres in Figure 3-1 shows that Flash B occurred *before* light sphere A reached B's location. Again, this is easiest to see by considering the direction from A toward B. The light sphere from B got a head start on the approaching light sphere A. Since the light-speed signal from A had not yet reached B when B flashed, then A could *not* have caused B to happen. These events are non-causal. Light spheres where neither sphere *completely* encloses the other will be called "non-nested."

Notice that causality, or non-causality, is connected to the nesting property of the light spheres: Nested means causal, barely nested means barely causal, while non-nested means non-causal.

As time passes, these three pairs of light spheres will grow ever larger, but their character (nested through non-nested) will forever remain unchanged. We will refer to this whole collection of conclusions about the light spheres as the "Light Spheres Theorem."

Notice that when we discussed the light spheres in Figure 3-1 we never did mention what frame of reference we were using for the description. Explicitly stating the reference frame has always been necessary before. Did we make an error in not specifying the frame? Since we never did need to define the reference frame to make the arguments about the light spheres, it must make no difference which frame we use for the arguments in establishing the Light Spheres Theorem. In another frame that is moving relative to whichever frame we used in Figure 3-1, the sources would have different motions, but Postulate 2 states that the speed of light does *not* depend on the motion of its source, so light spheres will be formed in any frame. The Light Spheres Theorem may seem to suggest that the same nesting character of the light spheres will be formed in all frames, thus all causality relationships would be the same in all frames. While this turns out to be true, we cannot justifiably make that claim yet. It is simply too big of a logical leap, and is not really all that believable to a person who is in his first exposure to relativity. Also, if we should stop here, we would miss discovering an important extra truth about causality across different reference frames, a simple result that can be very helpful when applying relativity to certain situations. We will return to the light spheres after we have established that relativity *is* compatible with causality. By looking at the light spheres when we did will give us better insight for the logical arguments that follow.

3-3 Showing That Causality is Preserved by Relativity

We will have the following setup for each of the next three gedanken experiments: 1) Choose any reference frame at all, and call it Frame 1. 2) Now suppose that there is a "sea of clocks" filling this frame. This is so that we will have a clock at any location that we will need. These clocks have all been synchronized in Frame 1 by the WWV method (see Appendix A8-3). These clocks have light detecting capabilities and will broadcast the time of arrival of any light pulse. 3) A flash occurs at some point that we will call "A." The clock at point A broadcasts the time of the flash. The flash from A reaches a completely arbitrary point, "B", that is *not* the same point as A. The clock at B broadcasts the time of arrival of the light pulse from A. The times between the emission and reception of the flash will be measured by subtracting the reading of clock A from that of clock B.

In the next three gedanken experiments we will use part of the Light Spheres Theorem without stressing it. This will be the portion that states that the light sphere produced by Event A will remain centered on the position of Event A at the moment of the event (flash) in whatever reference frame we are considering, and will not be affected by any motion of Event A.

We will begin by considering this situation (flash from A reaches B) from the point of view of another frame that is moving at a speed *v* along the line that joins Events A and B. This is shown in Figure 3-2.

The reader might find some of the following discussions a little challenging. It is recommended that the steps be taken slowly, trying to understand each. If completely understanding a step eludes your efforts, then try backing up a step or two, then trying again. If this does not help, then proceed forward a step or two to see if this helps.

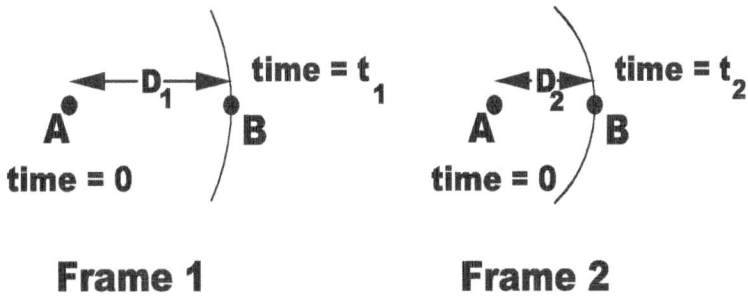

Figure 3-2. Frame 1 shows Event A, which emitted a flash and the light sphere is passing event B. The two events are separated by a distance D_1 and also by a time t_1. Frame 2 shows the very same events from the point of view of a frame that is moving parallel to the AB line. In this frame the events are separated by a distance D_2 and a time t_2, both of which are shown as smaller values than those in Frame 1.

Figure 3-2 is a one-dimensional (1-D) case since all motion (the light that reaches B and the observer in Frame 2) are moving along the line that joins the locations of Events A and B. After this simple one-dimensional case, we will broaden to a limited 3-D case, then finally show that relativity preserves causality in a general 3-D situation. We already have the tools to jump straight to the general 3-D case, but the approach that we take will provide a more gradual buildup and a better experience for the reader. In Frame 1, the two events are separated by a distance D_1 and by a time t_1. The events are barely causal, the light from the flash at A just reached B in the time t_1. Thus, it is clear that $D_1 = ct_1$. Frame 2 is moving to the right (or the left) at a speed, v, relative to Frame 1. In Frame 2 the spatial separation between the two events is contracted by the relativity factor R to $D_1 R$. Also, the time between events A and B will be dilated by the same factor to $t_1 R$. Take $D_1 = ct_1$ and multiply each side by R. This will give $D_1 R = c(t_1 R)$. This equation is simply $D_2 = ct_2$ (by recognizing D_2 and t_2). Thus, the two events will also be barely causal in Frame 2. Since we put no restrictions on what the value of v is (other than the usual limitation that its magnitude must be less than c, that is, a possible speed), then we have established that in this one-dimensional (1-D) case, causality is preserved by relativity. This argument is essentially the same one we used in Section 1-13 where we discovered length contraction.

Figure 3-3 shows the same pair of events, A and B, that are shown in Figure 3-2, Frame 1. Here we will consider the case where another observer, who is at rest in Frame 2, is moving at any possible speed v, *upward* in a direction that is *perpendicular* to the separation between events A and B (in Frame 1). In Frame 2 Event B has moved *downward* at a speed v, as required by Corollary 1. Any motion of Event A is immaterial by Postulate 2 and the Light Spheres Theorem. In Frame 1 the spatial separation between Events A and B, D_1, is a *transverse* length, (perpendicular to the line of relative motion between the two frames), and transverse lengths are absolute (the same in both frames), thus $y = D_1$. (see Section 1-12 if you are rusty on this point.)

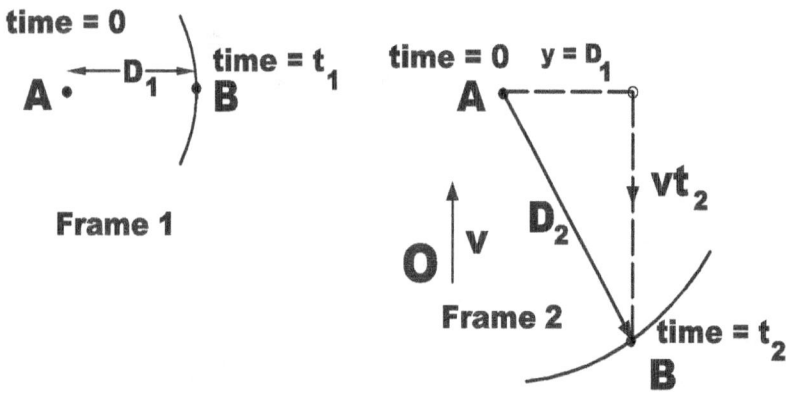

Figure 3-3. Frame 1 shows the same events as in Figure 3-2. Frame 2 shows these same events as determined by an observer, O, who is moving upward at a speed, v, perpendicular to the line of separation between Events A & B in Frame 1.

In Frame 2 the portion of the light sphere that travels between A and B will follow the diagonal path shown in Figure 3-3, Frame 2, and it will require a time t_2 to traverse this separation. At this point we do not know what t_2 is, so as usual we give it a name so that we can work with it. During the time t_2, Event B will have moved a distance vt_2 downward (in this upward-moving frame). In a 2-D or 3-D situation, the spatial separation (space) between the two events will usually have two components: One could be *parallel* to the relative motion between the two frames, and a second component would then be *perpendicular* to the relative motion between the two frames. See Figure A4-5.

To simplify our upcoming logical arguments, it is highly desirable to re-write the barely causal relationship, $D = ct$, by squaring both sides. This gives $D^2 = c^2 t^2$ which may be written $D^2 - c^2 t^2 = 0$. With the squaring we do not have to deal with the signs of D or t. We will refer to the left side of this equation as the "causality form." These last two equations give the causality question a link to the Light Spheres Theorem, and also a link to the Pythagorean theorem – see Appendices 4, A17-2, and A17-3. If the relationship between Events A and B is *non-causal*, then this means that the spatial separation, D, between the two events is too great for a light-speed signal to traverse this distance during the time separation between the two events, that is, $D > ct$. So, instead of a zero on the right side of $D^2 - c^2 t^2 = ?$, we will have some *positive* value. Conversely, if the two events *are causal*, not barely causal, then the right side will be a *negative* value because the distance, ct, that light travels in the time interval, t, will be *greater* than the spatial separation, D. If the events are

50

barely causal, then the causality form will equal zero. We will have need of this way of writing the causality condition (with the terms squared) both in the present case and also later when we consider the case where the motion of the second observer is in some completely arbitrary direction relative to the separation of the two events in Frame 1 and moving at any possible speed.

To consider the situation shown in Figure 3-3 we will begin with the causality form in Frame 2, then use some of what we already know to transform to Frame 1:

$$
\begin{aligned}
D_2{}^2 - c^2 t_2{}^2 \; &= \; D_1{}^2 + v^2 t_2{}^2 - c^2 t_2{}^2 \; = \; D_1{}^2 + \left(v^2 - c^2\right) t_2{}^2 \\
&= \; D_1{}^2 - c^2\left(1 - v^2/c^2\right) t_2{}^2 = D_1{}^2 - c^2 R^2 t_2{}^2 \; = \; D_1{}^2 - c^2 t_1{}^2
\end{aligned}
\tag{3-1}
$$

(Note: Here we are showing the math steps involved in the chapter of the book, rather than placing the steps in an appendix, because all but three of the steps involve relativity considerations, rather than being only math manipulations.) (First line above): As determined in Frame 2: Starting with the causality form; Applying the Pythagorean Theorem (Appendix A4-1) to the right triangle in Figure 3-3 gives us: $D_2{}^2 = D_1{}^2 + v^2 t_2{}^2$, insert this for $D_2{}^2$; factor $t_2{}^2$ from the two last terms in the middle expression in the first line (see Appendix A2-2); (second line above) Factor $-c^2$ from the ( ) term in line 1 (imagine $-c^2$ multiplied back in to see that this is correct) , reverse the order of the two terms in ( ) ; recognize the term in ( ) as the relativity factor squared; realizing that less time passes in Frame 1, or in any other frame moving relative to Frame 2 (us), we must multiply our time (squared), $t_2{}^2$, by R^2 (which is less than one) giving $R^2 t_2{}^2 = t_1{}^2$, which is the square of the amount of time between events A and B in Frame 1. Points A and B are closer together in Frame 1 than in Frame 2, so naturally less time will be required for a light signal to traverse their separation in Frame 1. This whole argument is very similar to Section 1-9 on the transverse light clock where we discovered the relativity factor and the relationship between times in different reference frames. The only difference is that, there, A and B were mirrors and the motion between the frames was shown as horizontal rather than vertical.

The last form of Equation 3-1 contains no velocity, v. The disappearance of v mathematically shows that the same causality relation holds for any possible relative speed (the only meaningful speeds) of frames as long as the relative motion is perpendicular to the separation of the two events in one of the frames. From Equation 3-1, we now see that the causality form, $D^2 - c^2 t^2$, will also have the same *value* in any frame, not merely the same sign (or be zero). More about this point a little later.

It often happens that when something is shown to be true in some direction, and that it is also true in any direction perpendicular to the first direction (as we did with the previous two gedanken experiments), then this will mean that the result must be true for directions in between. However, we cannot make that argument here! This is because the relativity factor depends on the total speed, v,

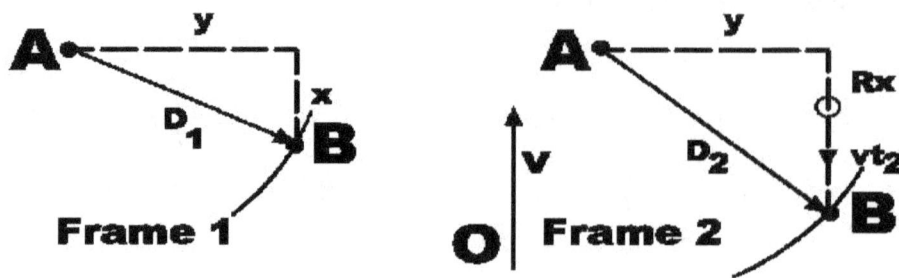

Figure 3-4.   Frame 1: Light leaves point A, the clock at A broadcasts its reading. The light reaches point B a time t_1 later. Frame 2: (The same events as determined by an observer, O, who is moving "upward" in Figure 3-4 at a speed v relative to Frame 1): Light leaves point A and reaches point B a time t_2 later. Note: In both parts of this figure the directions of the lines whose lengths are x and y have been drawn to be parallel and perpendicular to the direction of the motion of the observer, O. The page has been rotated so that the motion of observer O appears to be upward. None of this imposes any restrictions on the motion of observer O, but it does simplify our analysis and descriptions.

of one frame relative to another. The total speed will be a combination of the speeds in the two directions (see Appendix A4-1), where x and y are the two perpendicular directions. This will produce a mixing of the previous two results. As we will see in the next case, there will also be a clock asynchronization effect to account for that we did not need to consider in either of the two cases just analyzed.

For an all-inclusive (general case) analysis as our third and last demonstration that causality is preserved by relativity, consider the situation shown in Figure 3-4. To avoid errors we must be quite explicit about how the times are measured. We will be considering the situation from Frame 1 until further notice: As described above, light leaves point A at some time as indicated by a clock (not shown) located at point A. Clock A broadcasts this reading for all observers to receive. There is also a clock located at point B. These clocks are at rest in Frame 1 and have been previously synchronized in Frame 1 by the WWV method (Appendix 8-3). The clock at point B also has flash detecting equipment. When the flash from A reaches B, the clock at B broadcasts its reading for all observers to receive. The time between Events A and B in Frame 1, let us call it t_1, will simply be the reading of Clock B minus the reading of Clock A.

Still in Frame 1: The dashed line of length x is parallel to the relative motion between Frames 1 and 2. The dashed line of length y is perpendicular to this motion. The lengths x and y can have any values, including zeros and negative values. This means that the *direction* of the relative motion between the two frames can be in any direction whatsoever relative to the separation between the two events in Frame 1. The plane of the page in Figure 3-4 is the plane that contains the line of separation of Events A & B in Frame 1(the angled arrow showing the light path), and the line of motion between the two frames (the upward arrow). To define the plane of the page one of the two lines will probably have to be moved without changing its direction (This is called "translation".) until these two lines intersect. The intersecting lines will then produce an X or V shape which then defines the plane. If these two lines are parallel, then we need no translation to see that a plane is defined.

In Figure 3-4, Frame 2 is moving in a straight *upward* direction with a speed, v, relative to Frame 1. Thus, Frame 1 is moving *downward* with a speed v relative to Frame 2. The speed, v, can have any value that is less than c. Thus, Frames 1 and 2 can have any possible relative motion by adjusting the values of x, y and v. To an observer at rest in Frame 2, the motion of the clock at B will be downward. (Assuming a positive x, but this is not a limitation – the math will automatically take care of the situation.)

There are four relativistic considerations when we switch observers from Frame 1 to Frame 2: In Frame 2: 1) y is a transverse length (perpendicular to the relative motion between the two frames), thus it will have the same value in each frame. 2) x is a parallel (or longitudinal) length in a frame that is moving at a speed v. It will be contracted by the relativity factor to Rx as determined by the observer in Frame 2 (see Figure 3-4). 3) In Frame 2 the clocks in Frame 1 will be determined to be running slow by the relativity factor, R. 4) In Frame 2 the two clocks (at Points A and B) which were synchronized in Frame 1 will not be synchronized in Frame 2. Clock A is the "following clock," thus it reads ahead of Clock B by xv/c^2, because x is the separation between the two clocks in the direction of relative motion between the two frames in Frame 1 – the rest frame of the clocks. (See Equation 2-1, Exercise 2-2 in Section 2-2, and Appendix A8-2.) Again, the math will automatically take care of any plus or minus signs on v, x, or y. The effects of considerations 1 & 2 are shown in Figure 3-4.

Confusion can result in what we are about to do, so it is a good idea to consider an analogous and more familiar non-relativistic situation to obtain some needed practice. As an analogy, suppose that there is to be a bicycle race, and the course is a straight road between two cities in France. We (at rest in Frame 2) are the European Cycling Association and are responsible for accurate results. The race officials (at rest in Frame 1) have a clock at the starting line, and another at the finish line. The race officials will determine the time taken for a contestant to ride the course by subtracting the clock reading at the starting line from the clock reading at the finish line. However, unlike the race officials, we know that there are problems involving defective equipment. The clock at the starting line is mis-set so that it reads 5 minutes ahead of the clock at the finish line. We will correct for this error by adding 5 minutes to any reported time for a rider to complete the race. We have also learned that their clocks are defective and run at only $8/10 = 0.8 = 80\%$ of the correct rate (they run slowly). This is the same as if the race officials had *multiplied* correct time differences by 0.8. To make the necessary compensation, we must *divide* (the opposite of multiplication) our partially corrected times by 0.8 . Thus, to perform our entire correction we would take each of their reported time intervals, add 5 minutes, then divide by 0.8 . This would be:

$$Corrected\ time = \frac{(Incorrect\ time + 5\ Minutes)}{0.8} \qquad \text{(3-2)}$$

This is exactly the situation that we will face in our analysis of Figure 3-4 from the viewpoint of Frame 2 (us); only in our case the effects of unsynchronized and slow clocks are caused by relativity. We must keep in mind that we will be writing down an equation regarding Figure 3-4, from the viewpoint of Frame 2; we will refer to Frame 2 as "our" frame. (After any equations are written down about events in a frame, the equations can be faxed to any other frame for the manipulations such as those that are shown in Appendix 17. The rules of mathematics do not change from one frame to another. If the mathematical rules did change from one reference frame to another, then we would have detected absolute uniform motion – a violation of Postulate 1, which states that everything in any observer's frame must always seem completely normal.) t_1 will be calculated in Frame 1 and is the reading of Clock B minus the reading of Clock A. This is the same as the incorrect race time from the race officials. In our frame, Frame 2, Clock A is the following clock (see Figure 3-4, Frame 2), and therefore reads ahead of Clock B by xv/c^2 as explained in the previous example and is analogous to the 5 minutes in the bicycle race. As determined in Frame 2 (us), the clocks in Frame 1 run slow by the relativity factor, R, which is analogous to the 0.8 factor in the bicycle race. Thus, we have (paraphrasing Equation 3-2):

$$t_2 = \frac{t_1 + xv/c^2}{R} \qquad \text{(3-3)} \qquad \text{(in Frame 2)}$$

In this time, t_2 , event B will move downward a distance vt_2 , as shown in Figure 3-4. Applying the Pythagorean theorem (Appendix A4-1) to the triangle in Frame 2, we have:

$$D_2^{\ 2} = y^2 + (Rx + vt_2)^2 \qquad \text{(3-4)}$$

As seen earlier in this section, for causality we need to consider the numerical value of the form $D^2 - c^2t^2$ in each frame, where D is the space separation between two events, and t is the time separation between the same two events. To obtain this causality form for Frame 2 we simply subtract c^2t^2 from each side of Equation 3-4:

$$D_2^{\ 2} - c^2 t_2^{\ 2} = y^2 + (Rx + vt_2)^2 - c^2 t_2^{\ 2} \qquad \text{(3-5)}$$

We will now insert Equation 3-3 for t_2 into Equation 3-5 and then simplify. We note from Figure 3-4, Frame 1, and using the Pythagorean theorem, that $D_1^2 = x^2 + y^2$. These math-only manipulations are fairly long and are shown in Appendix A17-3. The result is:

$$D_2^2 - c^2 t_2^2 = D_1^2 - c^2 t_1^2 \qquad (3\text{-}6)$$

Notice that Equation 3-6 does not contain any of the variables, x, y, or v. The values of x and y depend on the direction of motion of the observer, O, and v is the speed of observer O relative to Frame 1. That these three variables cancel out of the equation means that their values make no difference. Thus, it makes no difference how Frame 2 moves relative to Frame 1. Equation 3-6 shows that the causality form has the same *value* in each frame. It then follows that the causal character (causal, barely causal, or non-causal) of two events will be the *same* in *any reference frame*. This is the result that we have been seeking for most of this chapter. When some entity maintains the same *value* in all reference frames, it is said to be "invariant." Frame 2 can be any frame, so from Equation 3-6, the causality form, $D^2 - c^2 t^2$, maintains its *value* in *all reference frames*, so it is invariant. The invariance of this form is a powerful and absolutely necessary conclusion. This invariance shows that relativity produces no problems for causality. Remember, causality must be correct; that is, every observer logically must determine the same causal character of two events, so our theory of relativity would have to be incorrect if it were incompatible with causality. [While we will not show it in this book, electric charge is an example of a physical quantity that is invariant. (maintains the same value from frame to frame).]

Equation 3-6 can also be useful in analyzing certain situations because of its equality, not merely that both sides are positive, zero, or negative together. For a simple example, return again to Section 1-9 and also Section 1-15 on the light clocks. If two observers have light clocks and both measure the separation between the mirrors in their own rest frames and obtain the same separation, that is, $D_1 = D_2$, then Equation 3-6 shows that if the mirror separations are the same, then each will measure the same time between ticks, $t_1 = t_2$, regardless of the orientation of the clocks in their frames, or even the relative motion of the two frames. However, Equation 3-6 is more general than this. For example, it also shows that if one observer's light clock has a larger separation between the mirrors than the other observer's, then the time between ticks will simply be proportionally longer, a completely normal (expected) result. This is all consistent with Postulate 1 (everything at rest in your own frame must seem perfectly normal). More complex problems may be simplified by exploiting the fact that for the same experiment, situation, or occurrence, the causality form has the same value in any reference frame. Equation 3-6 can be helpful even without considering causality. The D's must be the space separation between two events, and the t's must be the separation in time between the same two events. The subscripts (1 & 2) must refer to two separate reference frames.

For those few readers who are interested in a purely algebraic treatment to produce the same result – that relativity preserves causality in all cases – we need to use some powerful, but not too difficult, mathematics. This type of analysis does not lead to any additional intuitive understanding of relativity, but it is occasionally found in other books on relativity and is included in Appendix A17-1 and A17-2. There we use some of what we have already developed in Chapters 1 and 2 (contraction of longitudinal lengths, absoluteness of transverse lengths, time dilation, and the asynchronization of clocks) to derive a set of four equations called the "Lorentz transformation." From this transformation we can then show, by using only algebraic manipulation, that causality is preserved from frame to frame. While this proof is several pages long, it is no more mathematically involved than much of what we have already done, but those who still feel more than a little mathematically challenged may want to pass this part of Appendix 17 by for the moment. Appendix 17 does provide a link to other treatments of relativity that rely almost exclusively on the Lorentz transformation for their development. This is the primary reason for including this (optional) proof in the appendices.

There can be some confusion about what we have just done. One might say, "What if point B is moving in Frame 1? We didn't seem to take this possibility into account." If you thought of this, then you deserve a gold star. (again?) First of all, remember that any motion of the source of the flash at point A will make no difference in the light sphere that is produced, as explained earlier by both Postulate 2 and the Theorem of the Light Spheres. As to motion of point B other than that caused by the relative motion between Frames 1 & 2, we need to remember that two events happened: A flash occurred at point A, then that light reached point B. Considering the views of these *same two events* from two separate frames of reference is what we have been doing. It makes no difference how whatever else, if anything, that there might be at point B, got there. There need be only the clock and its associated flash detection equipment at point B.

We must not consider the events to have been a *ball* thrown from point A to point B. This is because relativistic causality depends on *possibly* causing. Using the fastest possible signal, for example, light, gives the greatest probability that two events will be relativistically causal. After all, by using a zero-speed signal, nothing could be causal. Mathematically, being causal means the causality form,

$D^2 - c^2 t^2$, is negative. Here, c is the speed of whatever signal we use. It is clear that a larger value of c will *tend* to make this form negative, that is, *causal*.

There is an alternate strategy and name that is used in deriving what we have called the "causality form." It is more often called the "space-time interval" (squared). If the spatial separation between the two events, D, is greater in magnitude than ct, the distance that light can travel in the time separation between the same two events, then the space-time interval is said to be "space-like." Oppositely, if the reverse is true, then the space-time interval is said to be "time like." It is more intuitive to consider the easily visualized "causality", so this is the strategy that we have followed.

3-4 The Light Spheres – Revisited

We will now return to the light spheres. Figure 3-5 shows what the light spheres in Frames 1 & 2 would be like for the barely-causal situation shown in Figure 3-4 (middle). While the larger light sphere is not the same size in the two frames, the two light spheres do have the same nesting character. There has been an apparent rotation, indicating a different direction for the light path from A to B between the two frames. Look at Figure 3-4 and convince yourself that the light path from A to B would be rotated clockwise because of the motion of Frame 2.

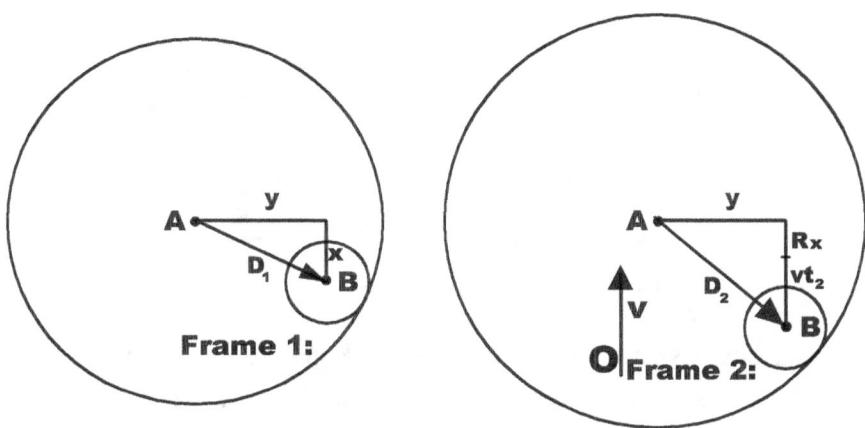

Figure 3-5. Light Spheres that form when Event B (in Figure 3-4) emits a flash immediately upon receiving the light pulse from Event A, that is, barely causal. The spheres in each frame are shown at equal times (in their respective frames) after Event B flashes, hence the same radii for the two smaller spheres. Frame 2 moves upward with a speed v relative to Frame 1. The same light moves at the same speed in each frame, but *not* necessarily in the same direction (see the arrows D_1 and D_2).

The important point is that both pairs of light spheres still indicate the same barely causal character in the two frames. This rotation of the direction of light travel is shown in both Figures 3-4 & 3-5. The light sphere from event A is larger in Frame 2 than in Frame 1. This indicates that both the separation in space, and the separation in time between the two events is greater in Frame 2. This will not always be the case. Whether these are larger or smaller depends on the relative direction from A to B to the direction of motion of Observer O.

3-5 A Causality Example

For a concrete, quantitative, but simple example of causality, we shall return to a previous gedanken experiment. Consider Figure 3-6 which is half of Figure 2-2 with some dimensions added. Since Ava and Bob concluded that Events 1 and 2 occurred in opposite orders, these events could not be causal. That is, neither event could have been caused by the other. Remember that in our frame (Figure 2-1) the events are simultaneous, and because the events are separated in space, neither could have caused the other because some time would be required even for a light-speed signal to traverse their separation.

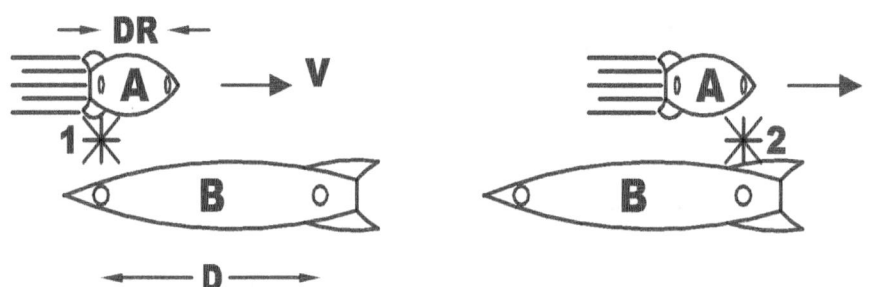

Figure 3-6. The passing identical ships from Figure 2-2 as seen in the rest frame of ship B (Bob's rest frame), shown with their lengths parallel.

Refer to Figure 3-6. In Bob's frame: It is simple to calculate the time to Bob, t, between when the left pair of windows align (and Flash 1 occurs) and when the right pair of windows align, causing Flash 2. If Ava in Ship A is traveling to the right at a speed v, then the time difference between the two images will simply be the distance that ship A moves in going from the left image to the right image divided by v (time = distance over speed). If ship B has a rest length of D between its windows, then in Bob's frame, identical, but moving ship A will have a contracted length of DR between its windows. Thus Ship A will have to move a distance of $D- DR$ to align the right windows, strike the second pair of matches and produce Flash 2. Using time = distance over speed, we obtain: $$t = (D- DR)/v$$

The pertinent question now is "Can a light-speed signal travel from the location of Flash 1 to Flash 2 in this amount of time?" That is, could Flash 1 have caused Flash 2? Remember, it need not have actually caused it, only that it could or could not have caused it. During this time t, light will travel a distance of ct. Thus, the question becomes, "How does ct compare to D, the separation between the flashes?" Multiplying both sides of the equation above by c, we obtain: (see Appendix A2-2 on factoring)

$$ct = \frac{c(D- DR)}{v} = \frac{Dc}{v}(1- R) = \frac{Dc}{v}\left(1- \sqrt{1-v^2/c^2}\right) \qquad (3-7)$$

58

Table 3-2 shows the values of ct calculated using Equation 3-7 with some representative values of v (the speed of Ship A relative to Ship B).

Table 3-2. Equation 3-7 for selected values of the speed, *v*, of Ship A relative to Ship B:

v (ship speed)	ct (the distance that light can travel in time t, via Eq. 3-7)	
0	0	(ct is less than D ⇒ non-causal)
$3c/5$	$1/3\ D$	(ct is less than D ⇒ non-causal)
$0.9c$	$0.63\ D$	(ct is less than D ⇒ non-causal)
$0.999c$	$0.956\ D$	(ct is less than D ⇒ non-causal)
c	D	(ct is equal to D ⇒ barely causal)

Everyone can use a calculator to plug in any value of v into the right side of Equation 3-7 that they like that is less than or equal to c to calculate the distance that the light signal can travel before Flash 2 occurs (other than $v = 0$; for this see Appendix 11). These representative values of v seem convincing that the light signal will *not* reach the right pair of windows before they line up for any speed v that is less than c. However, Table 3-2 does not *prove* this because it does not contain all possible speeds. (there could be an infinite number of possible speeds.) The last line shows that if the ship's speed is c, then a change in causality occurs. This is another logical reason why an observer cannot move at c.

Appendix A12-3 *does prove* that ct will be less than D, the spacing between the windows in a ship's rest frame, for any speed, v that is less than c, and that all steps in the logic are reversible, guaranteeing reversibility of the argument as it is used in the next paragraph.

What this gedanken experiment shows is that if v is less than c, then the right windows will line up before the light signal from Flash 1 can reach them. Thus, Flash 1 cannot be a cause of Flash 2, agreeing with (non)causality. (Remember, different observers disagreed on the order of Flash 1 and Flash 2.) With the argument in Appendix A12-2 being reversible, then we have also shown that *if causality is correct, then v must be less than c*. This is another strong reason to limit our study to speeds less than c, but we will have an even more compelling reason later.

One more statement follows from the arguments in the previous paragraphs: Since causality must be correct, then: If relativity is correct, no signal (information of any sort) can travel faster than c!

3-6 In a Limerick Frame of Mind

There was a young girl, Miss Bright
Who could travel much faster than light.
She departed one day
In an Einsteinian way,
And came back the previous night

There was a young man named Fisk
Whose fencing was exceedingly brisk
So swift was his action
That the Lorentz contraction
Reduced his rapier to a disk

Observe that for muons, created,
The dilation of time is related
To Einstein's insistence
Of shrunken-down distance
In the frame where decays aren't belated.

There was an old man named Lars
Who attended relativity seminars
So great his confusion
He made a resolution
And the presenter is now behind bars

The effects of dilation of time
Are magical, strange, and sublime.
In your frame, this verse,
Which you'll see is not terse,
Can be read in the same amount of
    time it takes someone else in
    another frame to read a similar sort of rhyme

The Muons and Dilation of Time are from David Morin,
www.physics.harvard.edu/limericks.htm Both are used by permission.

Miss Bright and Mr. Fisk are from the author's 1962 student notes. Origin unknown, but several versions exist at later dates. Miss Bright is clearly fantasy, since speeds cannot exceed c. Mr. Fisk is partially correct. Thrusting the rapier forward would cause a relativistic contraction in its length, but a much larger effect comes from causality. If the handle could be thrust forward extremely quickly, then the handle would move forward, but the tip would remain stationary until a signal could move from handle to tip. In reality, this signal speed would not be c, but rather the speed of sound in hard steel, a mere 6000 meters per second ($0.00002\ c$).

For the muons, one needs to think about which frame is being mentioned; it's the rest frame of the muons. The time dilation limerick is a clever joke, not to be taken seriously; everything at rest to an observer will always seem normal. All uncredited limericks in this book are by the author.

3-7 Two Classic Relativity Paradoxes

We are now in position to consider two of the most famous puzzles in relativity. The first is the "Stick and Slot Paradox" (Variations might be called the "Pole-Vaulter in the Barn," or something similar.): In their own rest frames there exist a meter-long stick and a meter-long slot. The stick, at rest relative to the slot, will barely fit into the slot. Now consider the stick moving parallel to its length and also parallel to the slot length. Of course, we could consider the slot to be the object

Figure 3-7

that is moving, only relative motion has any meaning (Postulate 1). In the rest frame of the slot, the stick is moving and is therefore contracted to a shorter length. Thus the contracted (moving) stick can easily fall into the slot. But, in the stick's rest frame, the stick is at rest and the slot is moving. In this frame the slot length, like all lengths parallel to the relative motion of the two frames, is contracted, but the stick is still one meter long. How can the stick now fall into the contracted slot? That the stick falls in the slot is *not* something about which observers in the two frames can disagree. Exercise: Resolve this paradox. At least come up with the mechanism. Hint: Think about causality. If one part of the stick or slot stops, will the remainder of the stick or slot stop all at once? Consider Mr. Fisk's rapier. Non-mathematicians should leave the quantitative work necessary to show that the solution really works to Appendix A12-2 .

The second puzzle is the famous "Twin Paradox": We have a pair of identical twins. *In Earth's frame*: The traveling twin, Roamer and his trusty companion, Rover, blast off from Earth and travel at nearly the speed of light to a distant star system. Because they are traveling so fast, time passes more slowly to them than to Homer, the stay-at-home twin, and also to Houser. On reaching the distant star, Roamer reverses course and heads back to Earth at the same very high speed – his time is still passing more slowly than back on Earth (in Earth's frame). When Roamer, the traveling twin, returns to Earth he finds that a lot of time has passed on Earth. His brother is an old man, yet he himself has aged only a little. The paradox arises when one considers the situation from the *frame of the traveling twin*. To him, it is Earth, with his twin, that moves away and back again at the same great speed (Corollary 1). Everything else seems the same, so Roamer incorrectly concludes that Homer should be the younger, not the older on their reunion. Part of the solution to this paradox lies in a careful reading of Postulate 1. It is absolute *uniform motion* (constant speed and direction) that cannot be detected. Accelerations can certainly be detected. Thus, there is no doubt as to which twin did the actual blasting off and turning around at the far end of the trip (accelerating at both ends of the trip). The remainder of the solution will have to wait until we have looked into gravity in Chapter 5. You may ask, "Gravity?" Yes, believe it or not.

Note on being younger: Roamer will not actually live a longer life than Homer. In Earth's frame the traveling twin's time *is* actually slower. All processes slow down. This includes clock speeds, thought processes, as well as aging processes. Roamer could not get more work done, read more books, or even watch more TV because of his motion (relative to Earth). Remember Postulate 1? Things at rest in your rest frame always seem normal. There may need to be more government forms to fill out and tax implications if we ever travel at speeds close to c. The Astronaut's employment contract should state his salary per year. But is it an Earth year, or a spaceship year? His age on Earth becomes complicated. His spaceship pilot's license might expire back on Earth without his knowing it. Roamer would be expected to die at a later Earth date than his twin. Unless his lease has a relativity clause, Roamer should definitely not keep his apartment rented while he is on the trip. (Unless he sublets it profitably and has an honest agent.)

Chapter 4 - Relativistic Momentum and Energy

We now move on to apply what we have learned about relativity to some real-world examples. In these examples moving from classical (pre-relativistic) thinking to relativistic thinking has a huge impact. Humanity's understanding of relativity has changed the world conditions in which we live!

Do not let this chapter title scare you. We will not be dealing with forces, friction, inclined planes, electric circuits or any concepts that require any special knowledge or experience. We will deal only with momentum and energy, and these two concepts are already at least partially understood by the public.

4-1 Momentum and Mass

Anyone who has watched or listened to a sports broadcast or followed a political election has heard that this team or candidate now has the momentum, or that team or candidate has lost the momentum. The idea seems to be that momentum is some kind of charging-ahead-ness. The notion also seems to be that once you have momentum you tend to keep it, unless something happens. Likewise, if you do not have momentum, then it takes some action to acquire it. This is about as good a description of momentum as we can get. A physical idea that has been around for centuries that has the above-mentioned characteristics is that momentum is equal to an object's mass, m, times its velocity, v. Momentum is usually given the symbol, p, so $p = mv$.

Notice that in the previous paragraph, we used the term "velocity" rather than "speed." Since an object can have not only motion, but motion in a certain direction, we need to differentiate between these two words. The term *velocity* means that we specify the motion with both speed and the direction of motion. For example, "We are driving at 70 miles per hour (the speed) relative to the road (specifying the reference frame) eastward (the direction of motion)." Velocity is an example of a *vector quantity*; that is, it has magnitude and direction. Other examples of vector quantities are: Displacement (50 feet up), and g (for gravity = 9.8 m/sec^2 toward the center of Earth). Try to think up some more yourself. (A downward moving bank balance is not a vector. We only have vectors in two or more dimensions.) We do something different with the wind vector. We give the direction from which the wind is coming, rather than going. A "north<u>erly</u>" wind is from the north, that is, a vector toward the south.

Since momentum is mass (which has no directional character) times velocity (which does), then momentum is a vector. Its units would be kg m / s (that is: kilogram meter per second). Notice that in the first paragraph, the symbols for both momentum and velocity were written in bold print. Vector quantities are usually bolded or have an arrow over them. Concepts without directional character, e.g., money, speed, energy or mass, are called "scalars" and are not bolded.

But what exactly is mass? An operational definition could be: Mass is that which is measured by a balance. While accurate, this is not a satisfying definition. Because mass is a property of matter that manifests itself in several ways, it is difficult to define. We will have to be content with a list of some of its properties and hope that the idea develops: Mass is attracted by gravity. This gravitational attraction is called "weight." Weight is proportional to mass; an object with twice the mass will have twice the weight, hence the confusion between mass and weight. One major difference between mass and weight is that weight is a vector quantity that, here on Earth, points toward the center of Earth, while mass is a scalar quantity (no directional character). The proportionality factor, *g* (for gravity), between weight and mass (***weight*** = *mg*) will be in different directions and have different magnitudes in different places. The same object (same mass) will weigh more on Earth than on the Moon, where gravity is weaker. On Earth, the same object will weigh very slightly less on a mountain top than at sea level because on the mountain the object and Earth are more separated, and gravity becomes weaker with increasing separation. However, an object will still have the same mass even if it is far from any object that would attract it gravitationally. As seen above, mass is one of the factors in an object's momentum. Relatedly, mass gives an object inertia (stay-put-ness or keep-going-ness). An object's mass will be nearly proportional to the combined number of protons and neutrons that it contains. (If you are not sure what protons and neutrons are, see the Glossary.)

A "law of nature" is a simple statement that makes a claim about something for which there is much experimental verification, and there is essentially no doubt of its validity. A law is less complex than a theory. Relativity is definitely a "theory" rather than a "law." The law that we presently need is the "law of conservation of momentum." By "conservation," we do NOT mean that the world is running low on momentum and that we should save some for future generations. What IS meant is that if any event occurs, and it is not being acted upon by some outside force such as gravity, then the *total* momentum (Appendix A4-1 tells how to add up vectors) of all the objects involved in the event will be exactly the same after the event as it was before the event. For example, suppose there is a collision between two objects. To have a collision, at least one of the objects must have been moving (have non-zero momentum). The two objects might simply bounce off each other, they might stick together, part of one might stick to the other, they might break into many pieces, they could explode (chemically or nuclear) emitting light and many pieces, or they might even completely miss each other. No matter what happens, the total momentum (the sum of all the momentums) of all the pieces before the event and the total momentum of all the pieces after the event will be identical. This is what is meant by "conservation of momentum." That is, we have neither gained not lost any momentum. No experiment has ever shown any reason to doubt its validity. Actually, an experiment once did look like momentum was not conserved in a decay (breakup) of some atomic particles. It was then predicted there must be some undetected particle, later called a "neutrino" [Italian for "neutral (uncharged), small"], that had the missing momentum. This was later confirmed. So, faith in the law of conservation of momentum resulted in the discovery of a new kind of particle.

One more thing about momentum before we can get back to relativity: Since momentum is a vector quantity, it has *components* (see Appendix 4-1) in each of the three spatial (space) directions. Each of the three components must be conserved, i.e., have the same value before and after an isolated event. An example from billiards or pool is that when the cue ball strikes a colored ball head on, the cue ball stops, but the colored ball moves in the same direction (momentum is a vector) and with the same speed as the incoming cue ball. When the cue ball strikes a colored ball off center, both balls move partially in the direction of the incoming cue ball and also have velocity components at right angles (90°) to the direction of the incoming cue ball. The law of conservation of momentum states that the sum of the two forward components of the balls will equal the incoming (forward) momentum of the cue ball. The transverse (at right angles to the incoming direction of the cue ball) motions of the two balls after the collision will be equal, but opposite in direction (they will add to zero), because before the collision there was zero momentum transverse to the incoming line of the cue ball.

4-2 Relativistic Momentum

Note of encouragement: You may think that you have no interest in relativistic momentum, but without the previous and this section we will not be able to arrive at the most famous equation in the world. (You already know what it is.) We are getting close, so hang in there. Prior to relativity we thought that momentum was given by $p = mv$. But will mv still be conserved when the relativistic effects such as slow clocks and shortened lengths are taken into account? It turns out not to be quite true. This old definition works extremely well for daily events on Earth, even space probes, but it will need a slight modification when speeds approach c. Since our usual definition works so well at low speeds, any new definition of momentum must give the same answers as the old definition when speeds are low. To discover the correct relativistic definition of momentum we will consider the situation (gedanken experiment) shown in Figure 4-1. Two spaceships pass each other with equal and opposite velocities as shown in the left half of Figure 4-1. Each ship has half of a batch of

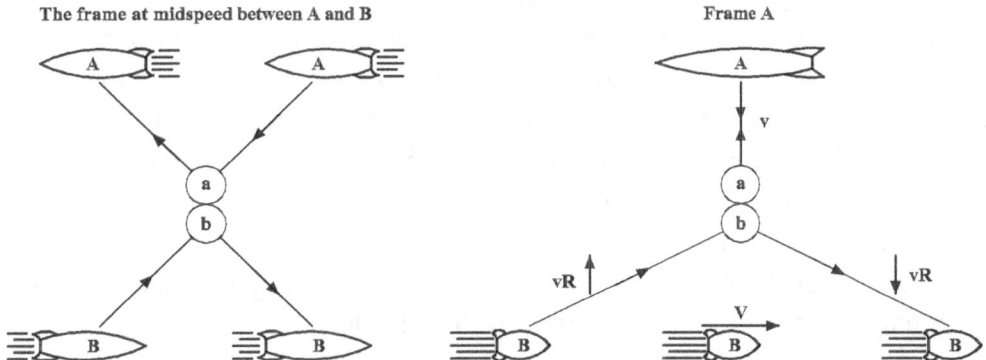

Figure 4-1 Identical perfectly bouncy balls are thrown straight out the windows of passing spaceships. The left figure shows the symmetrical situation as seen in the frame where the two spaceships are moving in opposite directions with equal speeds. The right figure shows the same situation, but in the frame where the upper spaceship is at rest.

identical, perfectly bouncy balls, each with a mass *m*. At the appropriate moment an astronaut (Ava or Bob) in each ship throws a ball with a speed *v* (as measured in the rest frame of his own ship), straight out the window (perpendicular to the line of motion of the two ships - see Ship A in the right half of Figure 4-1). Perfectly bouncy balls really do exist, helium atoms, for example. The balls collide and bounce back to the throwers. In the frame shown on the left the situation is completely symmetric. The vertical component of momentum of the balls is conserved. There was a total (both balls) of zero (equal, but opposite momentums) before the collision; there is a total of zero afterwards.

Now consider this same situation as it is **determined by Ava in Frame A** in which the upper spaceship is at rest. In this frame, ball *a* simply comes straight downward with a speed *v*, and returns straight upward with a speed *v*. Ball *b*, however, comes in even more angled than in the left figure in Figure 4-1.

According to Ava, Bob's ship is moving so his time will be running slow by the factor *R*, so Bob's arm will move slow by the factor *R*; thus, the vertical component of the velocity of ball *b* will also be reduced by the factor *R*. As seen in the left figure, the balls simply bounce back to the respective throwers as if they hit a mirror. It must be the same situation in the right figure. Logically, that the balls return to the throwers, or not, is something that both Ava and Bob must agree on. The fact that the balls return to the throwers cannot depend on the frame of reference (left or right view in Figure 4-1); they either do return or they don't. The astronauts could simply radio the information as to whether or not they caught the returning ball.

Let us assume that the balls are each thrown with a speed, *v* (in their own ship's frame), that is small compared to *c*. Let the speed of Frame B, that is, *V*, as measured by Ava, be very much larger than *v*. (We want relativistic effects to occur.) In that case, almost all of the speed of ball *b* (in Frame A) will be due to the ship B's motion, *V*. In Ava's frame, ball *b* appears to have lost some of its vertical speed, but it has gained horizontal speed, *V*.

At speeds much less than *c* momentum is quite accurately given by mass times velocity. Let us try a slight modification for great speeds by simply multiplying *mv* by a function, f(speed), which depends only on the speed. We will be trying the form: momentum = **p** = f(speed)×*mv*. To agree with the low-speed definition of momentum, the function f(speed) must approach the value of 1 at low speeds. Using this form for the *vertical* components of the momentums of the balls, and requiring that the vertical component of the total momentum be conserved, we must determine that the vertical components of the momentums of the two balls before the collision will be equal, but opposite. (We could use *after* the collision just as well.) That is:

To Ava in Frame A: (vertical components of momentum)

Ball *a* (down)	=	Ball *b* (up)
(speed of *a*) *m v*	=	f(speed of *b*) *m vR*
1 × *m v*	=	f(*V*) *m vR*

As argued in the previous paragraph, the function f is to approach the value 1 at low speeds (the speed of ball *a*). The total speed of ball *b* will approach *V* for large values of *V* and small values of v. Dividing both sides by mv, we see that the function of velocity, f(speed), must be $1/R$ (so that R will divide out), to conserve momentum. This function, $1/R$, is tabulated in Table 1-1 (Section 1-10). Now you know why we included it there. Notice that it does approach the value of 1 for small velocities, as required. Our relativistic expression (formula) for momentum is then:

$$p_x = \frac{mv_x}{R} = \frac{mv_x}{\sqrt{1 - v^2/c^2}} \qquad (4\text{-}1)$$

Remembering that momentum is a vector, the speed in the numerator, v_x, is one of the three velocity components (here we call it x, but 'vertical' in the above case), while the speed in the denominator, v, is the total speed because total speed is what appears in the relativity factor and affects time. The complete definition for momentum can be written as:

$$\boldsymbol{p} = m\boldsymbol{v}\Big/\sqrt{1 - v^2/c^2} \qquad (4\text{-}2)$$

(Vectors are written in bold.) This is the definition of momentum that is meaningful, and conserved in all cases and at all speeds.

Let us take a look at our new definition for momentum. Looking back at Table 1-1, it is easy to convince oneself that for any speeds of objects that we have dealt with, other than sub-atomic particles, the presence of R in the denominator has no measurable effects on momentum. This is why this factor was not discovered experimentally much earlier. Notice that neither simultaneity nor causality nor length contraction nor the non-synchronization of clocks entered our argument. It was simply time dilation (slowing, or stretching-out of time). Length contraction of the ships is shown in Figure 4-1, but we never used it.

What we have done here is not a proof because we have considered only one case. Another shortcoming of this analysis is that we did not consider the case where the astronauts throw the balls at speeds near c. However, our simplified analysis here does give the same expression that more complete and complex analyses give. An exact and more complete analysis is in Appendix 16 for those who are interested.

At very high speeds the relativistic expression for momentum gives much greater values for the momentum than the older expression gives. Verify this for yourself by looking at the $1/R$ column in Table 1-1. This column is the ratio of the relativistic momentum to the classical (pre-relativistic) momentum. At $v = c$ the relativistic (actual) momentum becomes infinite, but not the classical $p = mv$ ($= mc$ if $v = c$).

4-3 Relativistic Mass

In the previous expression for momentum some relativists like to group the "rest mass," m, (measured when the mass is *not* moving relative to the balance being used to measure the mass) with the square root, R, in the denominator and call the combination, m/R, the "relativistic mass." This would preserve the old definition of momentum as long as one remembers to use relativistic mass rather than the rest mass. Use of this concept is partially a matter of taste. One problem with the concept of relativistic mass is that while its use preserves the old expression for momentum, simple substitution into other classical (before relativity) expressions such as kinetic energy may not work at all. In our study we will keep m as the rest mass and put in the square root (or R) when it is needed. (We will deviate slightly when we discuss relativistic kinetic energy.)

You may have heard that objects moving at very great speeds are actually heavier (more massive) than when they are at rest. This is suggested by the relativistic expression for momentum. This increase in mass or weight will arise again later in another and more convincing way.

4-4 What is Energy?

Today most people know quite a bit about energy. They know that one has to pay for it, or at least pay for some apparatus to gather it, so it does cost money. Most things do, so cost does not do much to define what energy is, but cost does seem to imply that energy is something tangible, unlike love, charm or personality. Energy comes in so many different forms that a concise definition is all but impossible to give. Consult any dictionary for proof of this. It is better to first give some examples of its many forms, then we will attempt a vague definition.

If an object has motion, that is, a velocity (in some reference frame) then it has energy of motion in that frame. This is called "kinetic energy." If a cannon ball strikes a stone wall it might knock the wall over. However, this knocking over is because the cannon ball has momentum. Any object, either with mass (the cannon ball) or massless (a photon) that has kinetic energy will also have momentum, and vice versa, but these concepts are not the same. For speeds much less than c it has been known for centuries that the momentum of a mass m is given by the expression $m\mathbf{v}$, but its kinetic energy is given by $\frac{1}{2}mv^2$. Two differences appear immediately. Momentum is a vector quantity (See Section 4-1), that is, it has a directional character that is in the same direction as the velocity of the object. No form of energy has any directional character, but in addition to this difference, kinetic energy is proportional to the square of the velocity. For the same object, doubling its velocity will double its momentum, but it will quadruple its kinetic energy.

So, the momentum of the cannon ball provides the push that knocks over the wall, but the kinetic energy of the cannon ball will not only provide the kinetic energy of the tipping wall, but will also provide the energy that is required to break the chemical bonds that hold the various bricks, sand grains, and even individual molecules together. Everyone knows that energy is required if a hand saw is used to cut off a wooden board. Some of the energy is converted by friction into heat, but if the saw is sharp then most of the required energy is spent in tearing apart those chemical bonds (turning "biguns" into "littleuns"), that is, a thin slice of the board into sawdust. The same is true in trying to separate a burned-on spot from a cooking pan, or a spot of dried mud on the floor. Rubbing or scraping (motion, that is, kinetic energy) is usually required.

A form of energy related to kinetic energy is thermal energy. An object at a higher temperature will have greater average kinetic energy in its molecules. We must not think that a higher temperature *causes* higher molecular speeds. The average kinetic energy *is* what temperature *is*. The distinction between kinetic energy of an object and its thermal energy is that when we speak of kinetic energy we mean that the motion of all the molecules in the object is essentially organized, that is, motion in the same direction. A ball with kinetic energy will also have momentum. Contrasting this, with thermal energy the direction of motions of the molecules is random, or disorganized, and the kinetic energies of the individual molecules are spread over a wide range of energies, so a hot ball will not necessarily have any total momentum. Because of the disorganization of the kinetic energies and momentums of the molecules, only a fraction of thermal energy can be converted into some more organized form of energy , e.g, ordinary kinetic energy, chemical energy, or electrical energy. To paraphrase Oscar Levant's quotation about puns and humor, "Heat is the lowest form of energy." Contrasting this characteristic, any organized form of energy can be completely converted into thermal energy. Organized energy turning into thermal energy is the process that may eventually bring on the "Heat Death" or "Entropy Death" of the universe hundreds of billions of years in the future. At that time, almost all of the energy in the universe would have degraded into heat.

A completely different form of energy from kinetic or thermal energies is "potential energy." This is defined as the energy stored in some configuration. Some examples are: a boulder on the top of a mountain has a higher potential energy than that same boulder would have at a lower elevation. If the boulder should tumble down the mountain, then it is well known that it speeds up as it descends - its kinetic energy increases as its potential energy decreases. Potential energy can be converted into all of the other forms of energy. Chips and flakes of rock will be broken off the boulder and other rocks as the tumbling boulder collides with them. Some of the kinetic energy will be converted into thermal energy. This happens because the tumbling rock hits other things. The impact causes the molecules to vibrate more rapidly, and this is thermal energy. Some of the kinetic energy of the rock will be converted into another form of potential energy. In the case of the rock flakes or the sawdust above, when the molecules are separated (chipping), energy is required to separate them. If energy were not required to separate the microscopic pieces, then every solid or liquid object would immediately fall into a pile of unconnected molecules or atoms.

Chemical energy is a type of potential energy. Take the case of a mixture of hydrogen gas and oxygen gas. These will combine to form water; this combining tendency is because the hydrogen atoms and the oxygen atoms attract each other. But if the gases are simply mixed, they will not combine unless the temperature is rather high. This is because there is an "activation energy" that must be supplied to get over some form of energy hump. An example that is easy to understand is our rock on the mountain. It does not immediately roll down the mountain, because if it could, it would have already done so. The rock is probably in some sort of depression in the soil or between other rocks. To get it started down the slope to let it get close to the material farther down, we need to supply the energy to get the rock out of its hole; this is the activation energy. In the case of the hydrogen and the oxygen, the activation energy is supplied by thermal energy. This will slam molecules together, producing some distortions so the molecules can combine.

The energy in a charged battery is chemical (potential) energy. If the battery is not connected to some load, for example a light bulb or motor, the chemical reactions are prevented from happening because of electrical forces between the molecules. By connecting the battery to a load, some of the electrical charges move (this is electrical current), that is, change position. This decreases the restraining electrical fields and allows the chemical reactions to proceed at a much faster rate. The chemical reactions maintain the electric fields within the battery that try to move the charges. Before a lightning bolt, its energy is stored in electric fields within the cloud, between clouds, or between the cloud bottom and the ground. When the strike occurs, this potential energy is quickly converted into light, radio waves, thermal energy, and even some sound energy. Some energy might even be spent in blowing something apart on the ground.

One might wonder about this potential energy. Where is it? How is it stored? To understand this one needs to consider "fields." We have all heard of gravitational fields, electric fields and magnetic fields. If any of these fields occupies a volume of space, then energy is present. That is, a certain amount of energy per volume is required just to make the field. This is called an "energy density" (energy per volume). For any type of field the energy density is proportional to the square of the field strength. If we lifted the tumbled rock mentioned above back up to its starting height, this would very slightly alter the strength and shape of Earth's gravitational field in the vicinity of the rock. This altered gravitational field is where the potential energy is stored. We can give a crude numerical example to illustrate this phenomenon: Suppose that Earth's gravitational field near its surface is 10 units. Also suppose that at a small distance from the rock the rock produces a gravitational field of 1 unit toward itself. Above the rock the combined gravitational field will be 11 units because both Earth's and the rock's gravity point in the same direction - downward. However, below the rock Earth's gravity is still pulling downward, but the rock's gravity is pulling upward (toward the rock). This will give a combined gravitational field of 9 units. But now the squaring of the field strength enters: 11 squared is 121, while 9 squared is 81. The Earth's field strength of 10, when squared, gives 100. Notice that above the rock its presence has increased the energy density by 21, while below the rock the energy density has decreased, but only by 19. By summing up the energy density all around the rock the net effect will be an increase in the energy stored in the gravitational field. Introducing complexity into the field shape and strength will always increase the energy in the field. Similar arguments can be made about any other type of field. Chemical energy is stored primarily in the electric fields produced by the charged particles that make up atoms and molecules. A lesser amount of energy is stored in microscopic magnetic fields.

A little thought on the subject of energy will reveal the wide scope of the concept. We can try to define energy as "the capacity to do something physical." These physical things include ripping things up, lifting something in a gravitational field, or increasing the speed, hence the kinetic energy, of something. The physical concept of energy is even used in a psychological sense. For example, "He is very energetic." In our study here, we will be referring only to the physical meanings of energy.

4-5 Mass and Energy in Relativity - That Most Famous Equation

We will now consider another gedanken experiment and arrive at that most famous equation in the world. We will consider the decay (a particle splitting into two or more smaller particles) of a particle with rest mass, M. In the reference frame where the particle is at rest before the decay, there was no movement, hence zero kinetic energy. After the decay there is kinetic energy. (If the particles do not separate, there is no decay.) Thus, the question arises, "Where did this kinetic energy come from?" The primary mission of this section is to answer this question. It should be realized that spontaneous particle decays like we describe in Figure 4-2 actually do occur in nature (well, in high energy accelerator experiments and the previously mentioned muons produced by cosmic rays - Section 1-16). For example, a K^0 particle (an electrically neutral K-particle, the superscript indicates the charge) will decay into two identical π^0 particles (neutral pions). We are taking the case where an object (particle) of rest mass M (see Sections 4-1 & 4-3) decays into two identical particles each of rest mass m. We WILL NOT assume that $M = 2m$ as we would normally think! We will soon find out if this is true.

We will now consider the conservation of the horizontal component of momentum as determined by Ava. Since the particle of rest mass M is not moving in Frame A, it has zero momentum in Frame A. Because momentum must be conserved, then after the decay there must still be zero total momentum because no external forces affected the process. If two identical particles (same rest mass) are produced in the decay, then to conserve the zero momentum from before the decay the only possibility is that the two particles have equal but opposite momentums, so the momentums will add to zero, which means equal and opposite velocities (in Frame A). We will arrive at this conclusion using either the old or the relativistic expression for momentum.

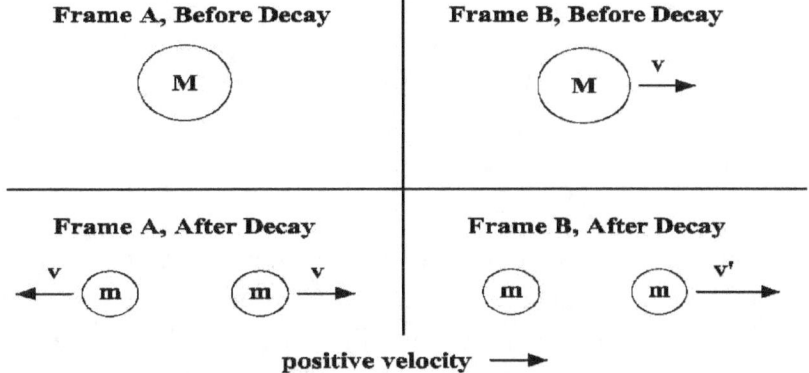

Figure 4-2 The decay of a particle with a rest mass of M into two identical particles, each of rest mass, m, as seen in two different frames. These show the **same event** from the perspectives of two different frames of reference. Frame **A** is at rest relative to the original particle *before* the decay and Frame **B** is at rest relative to the left decay particle *after* the decay. The vertical and horizontal lines separate four views: The decay process is shown as it is before and after the decay in each of the two frames. For our discussion, we have chosen the direction to the right as the positive direction of velocities.

We will now further examine this very same spontaneous decay event as determined by Bob who is at rest relative to the left particle of rest mass m <u>after</u> the decay. This clever choice for Frame B greatly simplifies our analysis, but we would eventually arrive at the same final conclusion with the choice of any frame that is not simply Frame A again. To Bob the initial particle of rest mass M (before the decay) will be moving to the right with a speed v which is equal to the speed of the left-moving and left particle of mass m in Frame A. These are reciprocal velocities. Corollary 1 (Section 1-6) assures us that these must be equal, but opposite velocities. (We can even get this result by using Equation 2-6, the velocity combining equation, as we do just below, but with $v = 0$ for the velocity of mass M in Frame A. The first method is easier.)

We need to calculate the speed v' that the right particle of rest mass m moves relative to the stationary particle of mass m in Frame B. Remember, it is NOT simply $2v$; we must use our relativistic velocity combining equation (Equation 2-6):

For the right decay particle of rest mass m:

v' is the particle velocity to Bob, v is the particle velocity to Ava.

$$v' = \frac{v - V}{1 - \dfrac{Vv}{c^2}} \qquad \text{(2-6 repeated)}$$

Where V (in Equation 2-6) is the velocity of Bob (Frame B) relative to Ava which is equal to $-v$ since Bob is moving to the *left* relative to Ava (the left decay particle is at rest to Bob); and v (in Equation 2-6), the velocity of the right decay particle with mass m to Ava, is simply equal to $+v$ since this is the speed (and direction) of the right decay particle in Frame A.

We obtain from Equation 2-6:

$$v' = \frac{v - (-v)}{1 - \dfrac{(-v)v}{c^2}} = \frac{2v}{1 + v^2/c^2} \qquad \text{(4-3, in Frame B)}$$

Notice that if v is a small velocity compared to c, then the second term in the denominator of Equation 4-3 will be very small, meaning that the denominator is nearly equal to 1. So at low speeds, v' would almost be simply $2v$. At higher speeds, v' will be less than $2v$. See Table 2-1 in Section 23.

Refer again to Figure 4-2. We will now apply the law of conservation of momentum in Frame B: Before the decay, only mass M has any momentum, and that is positive (to the right). After the decay, the left decay particle has zero momentum, but the right decay particle has a positive (to the right) momentum. Conserving momentum (using our relativistic expression for momentum - Equation 4-1):

Total Momentum before decay $=$ Total Momentum after decay (all in Frame B)

$\quad\quad\quad$ mass M left $+$ right decay particles (each of rest mass m)

$$\frac{Mv}{\sqrt{1-v^2/c^2}} = 0 + \frac{m\dfrac{2v}{1+v^2/c^2}}{\sqrt{1-v'^2/c^2}} \qquad (4\text{-}4)$$

where we have plugged in the value of v' from Equation 4-3 into the numerator of the right side of the expression of relativistic momentum (Equation 4-1), but not yet into its denominator. The algebraic manipulation that follows is several steps long; it is presented in Appendix A13-1 where it is shown that Equation 4-4 simplifies to:

$$2m = M\sqrt{1-v^2/c^2} \qquad (4\text{-}5)$$

Since the square root (R) is less than one for all non-zero speeds, Equation 4-5 shows that $2m$ will be less than M. This means that MASS HAS DISAPPEARED in this decay process! However, something else has appeared, namely *kinetic energy*. That is, the energy of motion of the two decay particles has appeared. (Kinetic energy can be completely transformed into other types of energy as was discussed at length in Section 4-4.)

This type of calculation and reasoning led Einstein to conclude that mass and energy are really two forms of the same thing. We call this combination "mass-energy."

Before relativity we had separate conservation laws: Mass conservation and Energy conservation. That is, it was then thought that each was always conserved, that neither could be created nor destroyed. They *could* be moved around and energy could change forms. Now, we have a combined law, "conservation of mass-energy." In principle, either mass or energy could be transformed into the other. (Here comes The Bomb?)

The question arises: "How much mass is equivalent to how much energy?" It seems reasonable to assume that the energy would be directly proportional to the amount of mass, not to the shape or color, etc. of the mass, nor to the amount of mass squared, where the mass is located, nor any similar thing. After all, no such suggestions arose when we derived Equation 4-5. The universe seems to operate on the "KISS principle." (Keep It Simple, Stupid!) We thus try:

$$E = km$$

Where E is the amount of energy equivalent to some mass m (NOT limited to the mass of one of the decay particles considered above), and the constant k is the proportionality factor. We need to find what k is. To do so, we will again return to the decay process as seen by Ava. We will apply our new law, conservation of mass-energy. From Equation 4-5, we obtain $M = 2m\Big/\sqrt{1-v^2/c^2}$ by simply dividing both sides by the relativity factor .

Expressing mass as its energy equivalent (k times mass) and including the kinetic energies:

Energy before decay　　　　　=　Energy after decay　(in Frame A)

$$(kM=)\quad k\,\frac{2m}{\sqrt{1-v^2/c^2}} = 2km + 2KE \qquad\qquad (4\text{-}6)$$

Before the decay all the energy was kM . That is, all the energy was in the mass of the initial particle with zero kinetic energy because the original particle was at rest in Frame A. But after the decay the energy is more spread around: It is in the masses of the two decay particles, each with km, plus the kinetic energies of the two decay particles (with equal speeds in Frame A).

For about 200 years before 1905 it was known that the kinetic energy for speeds that were attainable then was given by $KE = \frac{1}{2}\,mv^2$. We have yet to discover what the relativistic expression for kinetic energy is, but since all that we are trying to determine is the <u>constant</u> k, we may restrict our reasoning to just one (non-zero) speed. Let us choose some slow speed where the old definition of kinetic energy is known to be valid. As shown in Appendix 11, $1/R$ can be approximated by

$$1/R = \Big(1 + \tfrac{1}{2}v^2/c^2\Big)$$

for low speeds. Inserting these two low-speed forms into Equation 4-6 we obtain (the denominator on the left side of Equation 4-6 is the relativity factor, R) :

$$k\,2m\Big(1 + \tfrac{1}{2}v^2/c^2\Big) \;=\; 2(km) + 2\Big(\tfrac{1}{2}\,mv^2\Big) \qquad\qquad \text{(low speeds)}$$

Dividing both sides (all three terms) by m, multiplying k through on the left side, and realizing that 2 times ½ equals 1, we obtain:

$$2k + kv^2/c^2 \;=\; 2k + v^2$$

Subtract $2k$ from each side of the equation, then divide out the v^2 (if v were zero there would be no separation between the two decay particles, i.e., no decay); and we are left with $k/c^2 = 1$; multiplying each side by c^2, we get $k = c^2$. Thus, from $E = km\ (= mk)$, we arrive at:

Ta Dah!

$$E = mc^2 \qquad\qquad (4\text{-}7)$$

Here E represents the energy that is equivalent to some mass, m. c is a huge number, so c^2 is extremely large. This means that a small amount of mass has a huge energy equivalent. If we could turn about 100 pounds (45 kilograms) of anything, even the lowest form of matter scooped up in the barnyard, completely into energy we would be able to supply the energy needs of the United States for a whole year. This most famous equation just tells us that the energy is there, but it in no way tells us how to do the conversion of mass into energy, or vice versa. Fortunately, it is very difficult to do the conversion of mass into energy except on microscopic scales. This equation does NOT tell us how to build nuclear weapons! After all, this equation was known for nearly forty years before nuclear bombs were developed.

This most famous equation in the world looks very simple. One might have thought that it would spring from rather simple reasoning. On the other hand, it is sometimes referred to as "Einstein's equation." In many people's minds this might give the impression that it was very difficult and took a mega-genius to discover. The actual truth is somewhere in between. If the reader looks back at what all was involved in deriving this famous result, one will discover that everything that we have done, with the sole exception of the chapter on causality, has been used in the steps leading toward this equation. If the reader has also gone through the algebraic manipulations in the appendices (which have nothing directly to do with relativity), it is even more impressive that all that has been done condenses to such a simple-looking equation. Nature has a way of doing this, so the simplicity of $E = m c^2$ gives it the ring of truth. Of course, the "ring" does not prove anything. Part of the reasoning for the topics that have been included so far was to be able to derive this result. So famous and important is this result, that any treatment of relativity that does not *derive* this equation from the postulates, and considering any possible speed, v, must be considered as incomplete.

We need to do a few more things in this section. First, by returning to Equation 4-6, plugging in $k = c^2$ and dividing each side by 2, we obtain:

$$\frac{mc^2}{\sqrt{1-v^2/c^2}} = mc^2 + KE$$

Solving for the Kinetic Energy (by subtracting mc^2 from each side and turning the equation around):

$$KE = \frac{mc^2}{\sqrt{1-v^2/c^2}} - mc^2 = mc^2 \left\{ \frac{1}{\sqrt{1-v^2/c^2}} - 1 \right\}$$

$$= mc^2 \left\{ \frac{1}{R} - 1 \right\} = \left\{ \frac{m}{R} - m \right\} c^2 \qquad (4\text{-}8)$$

In the top line above we factored mc^2 out of the two terms, then we recognized the relativity factor. Equation 4-8 is the relativistic expression for the kinetic energy of an object with rest mass m moving with any speed v. In the second line above we multiplied m through each term and rearranged the product. This shows that the kinetic energy is the relativistic mass minus the rest mass all multiplied by c^2. Since the speed of an object varies depending on which frame of reference one is using, so does the kinetic energy vary with the choice of reference frame. This is true even without relativity. The kinetic energy of an object will always be zero in that object's rest frame. Exercise: Verify that Equation 4-8 does approach $KE = \frac{1}{2}\,mv^2$ at low speeds by simply inserting our low-speed form for $1/R$ from Appendix 11 into Equation 4-8. The solution is in Appendix 18. At least have a look at it.

We will obtain an interesting result if we add the rest-mass energy to the kinetic energy of the particle. This sum will then be defined as the "total energy" of the particle. We will temporarily call the total energy, E_T. Let us add Equations (4-7) and (4-8):

$$E = mc^2 \qquad\qquad \text{(4-7, repeated)}$$

$$(\textit{plus}) \qquad KE = mc^2\left(\frac{1}{R} - 1\right) \qquad\qquad \text{(4-8, repeated)}$$

$$(\textit{equals}) \quad E_T \equiv E + KE = \frac{mc^2}{R} = \frac{m}{R}c^2 \qquad\qquad \text{(4-9)}$$

So, from Equation 4-9, the total energy of an object equals its relativistic mass (Section 4-3) times c^2. The total energy is often written simply as E, and the relativistic mass is often written simply as m. Thus the equation $E = mc^2$ can be used in two ways: If m is the rest mass of an object, then E is the energy equivalent of this rest mass. (As discussed below, other forms of energy, such as thermal energy, can add to an object's rest mass.) If m is the *relativistic mass*, then E is the *total energy* of the object (rest-mass energy plus kinetic energy). This is further developed in Appendix A13-2 for those readers who are interested.

We close out this section by presenting a few important considerations regarding what we have derived: Returning to the section on Relativistic Mass (Section 4-3), Equation 4-8 can be seen to be the relativistic mass minus the rest mass, all multiplied by c^2. **Thus, kinetic energy does add to the mass of an object** (as shown in Equation 4-9). Moving masses really are heavier than when they are at rest. All forms of energy contribute to an increase in mass. An object is even slightly heavier when it is hotter. This *seems* to be measurable on a sensitive balance, but the real (relativistic) increase in mass is very tiny and the increased balance indication is actually caused by the surrounding warmed air rising (upward momentum), producing a slight downward reaction thrust on the balance pan. (Analytical chemists are probably aware of this effect.) This effect is not seen when the balance and warmed mass are in a vacuum where there is no air to produce the thrust.

Looking at the relativistic expression for kinetic energy, we can see yet another reason why objects with mass (not photons) cannot reach the speed of light. The kinetic energy would approach infinity as the speed approaches c. It is impossible to obtain an infinite amount of energy, not even considering the cost involved. Representative values of the classical and the relativistic kinetic energies are shown in Table 4-1 . (See Appendix A1-4 for an explanation of the scientific numbers.)

Table 4-1 Kinetic Energies (in joules) of a 100 kg mass (~ 220 lbs) at various speeds			
v	Classical KE	Relativistic KE	Ratio
0	0	0	1
100 km/hr (62 mph)	3.8580×10^4	Classical $+ 2.5 \times 10^{-10}$	~1
20 km/sec	2.000000×10^{10}	Classical $+ 66.7$	~1
$0.01c$	4.500×10^{14}	4.50036×10^{14}	1.00008
$0.1c$	4.500×10^{16}	4.53403×10^{16}	1.00756
$0.25c$	2.8125×10^{17}	2.9516×10^{17}	1.04944
$0.5c$	1.125×10^{18}	1.3923×10^{18}	1.2376
$0.75c$	2.531×10^{18}	4.6067×10^{18}	1.8199
$0.9c$	3.645×10^{18}	1.1647×10^{19}	3.1955
$0.99c$	4.41×10^{18}	5.4800×10^{19}	12.413
$0.9999c$	4.499×10^{18}	6.2741×10^{20}	139.45
c	4.500×10^{18}	∞	∞

One joule is the energy obtained when 1 watt of power is applied for one second. For a speed of zero, both the classical and the relativistic expressions for kinetic energy give the same result for kinetic energy, namely zero. At a typical automobile speed, there is only about a one part increase in 100 trillion by using the relativistic expression. 20 km/sec is about the speed of the fastest space probe we have ever launched. At this speed the relativistic expression is about seven parts per billion greater than the classical result. At 10% of the speed of light, there is almost a 1% increase in kinetic energy over the classical result. However, when speeds get really great, the increase becomes dominate and even goes to infinity at a speed of c. Again, this is the strongest argument that we can make about why speeds for objects with a rest mass cannot reach the speed of light. An infinite amount of energy cannot be obtained!

In case you wondered what would happen to that meter stick that fell into the hole and stopped suddenly (Section 3-7): A wooden meter stick moving at 10% of c would have a kinetic energy of about 5×10^{13} joules, which is also said to be 50 kilotons or 0.05 megaton (50 thousand tons of TNT - wow!). This is more than the combined energy output of both of the atomic bombs that ended WWII. Everything for miles around would be destroyed. You wouldn't want to be around to see that meter stick stop so suddenly, nor would you want to pay for all that energy to accelerate the stick to such a great speed. [One megaton (one million tons of TNT) will release approximately

10^{15} joules of energy. This is typical of the energy released by a hydrogen bomb, and the mass converted into energy is only 11 grams, or 2.8 onces.]

4-6 Interstellar Travel

How much energy would it take to accelerate a 150 pound person, no space suit, no spaceship, no supplies (peanuts nor pretzels) to 99% of the speed of light? The mass of this person (if she were weighed on the surface of Earth) would be 68 kilograms. This gives a kinetic energy [by Equation 4-8)] of 3.7×10^{19} joules, which is the energy released by 37,000 one megaton hydrogen bombs! This is about one tenth of the energy used by all persons on Earth (and their activities) in a year (in 2005). If she desires to travel at 99.9% of c, the energy required is more than three times as great. If she traveled on the economy plan (only traveling at 99% c in Earth's rest frame), then R would be about 7 / 50, so her time would be running slow by this factor to observers on Earth. Thus, she could travel about 500 light years (as measured on Earth) in 70 years , as measured on her watch. She would determine that her time runs normally, but the distance would have shrunk from 500 to 70 light-years. It would require a very large amount of extra energy just to let her take her watch along, much less her cell phone and notebook computer. Actually, after 70 years of ship time, the astronaut would probably not be interested in starting a return trip back to Earth - a huge energy and cost saving. Carrying the energy resources required for the return trip would increase the energy requirements upon leaving Earth by a huge factor, not just double it.

500 light-years is about the expected distance to the next civilization that is as advanced as we are, as calculated using fairly optimistic estimates of all factors that enter into such a guess. Realizing that a spaceship is needed, and when we include the mass equivalent of the energy that must be taken along to both accelerate it to this speed and then stop it at the other end of the trip, the energy requirements run into trillions of times the above-calculated amount. At the great speed of the ship, running into photons, protons, and electrons in space would produce x-rays that would require heavy shielding to protect the astronaut. Colliding with these particles will also cause drag on the ship, slowing it. All of these effects further compound the energy requirements. Perhaps this is a reason why there is no credible evidence of earthly visits from alien beings. Who would expend so much just to visit less-advanced Earth? Also, aliens at this distance would not yet even know of our civilization. Radio and light signals from Earth have had only enough time to travel about one-fifth of that 500 light-year distance.

It has been suggested that perhaps interstellar travel might be possible by using an "interstellar ramjet." This is a proposed spaceship that would scoop up the very thin gas (mostly hydrogen) that pervades the near-vacuum between stars. The forward motion of the ship would compress (ram) this gas into a reaction chamber where the hydrogen-bomb reaction (Section 4-7) could then produce so much energy that the exhaust gas would be expelled at relativistic speeds, producing quite a large forward thrust. Some large funnel-like mechanism would have to act as the scoop. Electromagnetic fields might do the scooping job, but some analyses of all of these scoop processes indicate that the ship would be unstable; the ship might try to turn around end for end, like a weather vane in the wind, making the whole process unworkable .

Another suggestion that has been put forward is the idea of using a super laser on Earth (where energy is "cheap") to push the ship along. Twice the push can be attained if the ship uses a mirror to reflect the laser light backward in the general direction of the solar system. The technical problems with this plan are mainly being able to build such a laser and the cost of operating it. There are also political problems. A few (Earth) years after launch, there might be a change in political administrations. The new administration might no longer continue the funding of the project. A further problem is getting the ship turned around at the destination and then back to Earth. Will the aliens that the ship is to visit build a laser to stop the ship at their planet and then later send it back toward Earth at their expense? On Earth, there would still be the energy expense of stopping the ship on its return. With such considerations, any interstellar ship should be self sufficient - an extremely expensive option.

If we use a spaceship that attains only modest speeds, then it will take hundreds or even thousands of years to make a trip to only the nearest stars. At a speed of 0.1 c (18,600 miles /second - about 1000 times the speed and a million times the kinetic energy of any spacecraft that has left Earth), it would take about 42 years of Earth time, and almost as much ship time just to reach Alpha Centauri, our nearest neighbor star. If we want to make a longer trip to a more interesting destination, such as a star cluster where there would be many stars to explore, then the time could get really long. If technology keeps advancing at the rate that it is in the very early 21st century, then before Ship X-1 could reach the destination, it might be passed by Ship X-2, a later and much faster design. See Figure

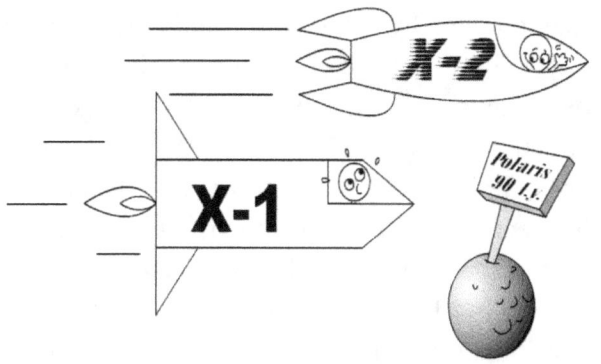

Figure 4-3

4-3. Then Ship X-1 would probably lose funding and ground support. This is a joke, of course, but the truth is that for a really long trip, with technology advancing, it might always be better to wait in order to get there sooner for less money. We currently (2005) have a similar situation with computer purchases. Waiting a few months will get you a faster computer cheaper.

Does relativity, with its speed limit of c, put limits on interstellar travel, or does it make it easier? This is an impossible question to answer experimentally because the universe is the way that it is; relativity and c are what we've got. We can get either answer if we consider only certain aspects of relativity. Probably the best way to attempt to find the answer is to imagine that c were a much larger number. This would greatly reduce all of the relativistic effects that we have seen, including the slowing of time that allowed the astronaut mentioned above to travel 500 light years (in Earth's frame) in 70 years of ship time. If c were thousands of times greater than it is, then the astronaut would have to travel about 50 / 7 times as fast to get there in 70 years of ship time. If c

were infinite, then all relativistic effects would vanish. For example, the relativity factor would always equal unity (one), and any fraction where c is in the denominator would be zero. Look at Equations 1-2, 2-1, 2-5 and 4-5. Using the classical expression for kinetic energy ($\frac{1}{2} mv^2$) we would find that about four times as much energy is required for our classical example than in our relativistic case. At this point in the discussion it appears that relativity actually makes interstellar travel easier. One might argue that with c so much larger, then $E = mc^2$ would allow us to extract even more energy from matter. However, with this larger value of c more energy would have been required to produce the matter in the universe. Perhaps we would have almost no matter in the universe at all, only energy. Since we and the solar system are made of matter, this is not an exciting idea. In any case we cannot change the laws of nature, but it is interesting to speculate.

4-7 Solar Energy and H-Bombs

This section, and the next, present applications of $E = mc^2$. Many people find the energy source for the sun to be quite interesting, and these sections point out one vital principle that must be considered when dealing with this famous equation.

Examine the four hydrogen atoms in Figure 4-4. Hydrogen atoms are the simplest of all. They are also by far the most numerous in the universe. Each hydrogen nucleus consists of a single proton. Protons have one unit of positive charge as shown. A single electron (negative charge) orbits the nucleus. The attractive force that holds the electron in orbit is not gravity, but rather the electrical attraction between unlike-sign electrical charges. There is a gravitational attraction between the proton and the electron, but it is less than one billionth trillionth trillionth (10^{-33} times) as strong as the electrical attraction. At the right is a single helium atom. It has two protons in its nucleus and two electrons in orbit. The helium nucleus also contains two neutrons. Neutrons are very similar to protons but have no charge. Almost all of the mass of any atom is in the nucleus; the electrons contribute less than one part in 1800 to the mass of any atom.

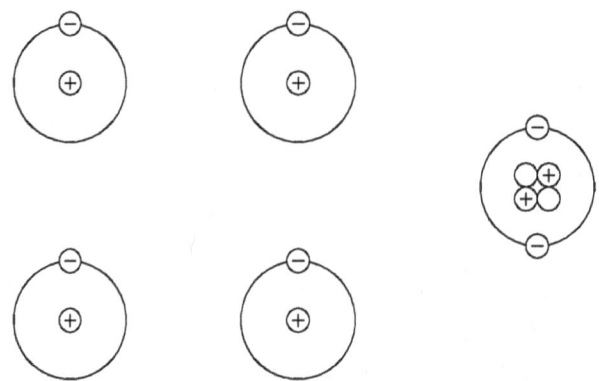

Figure 4-4 Four hydrogen atoms (left) have just the needed particles to form one helium atom (right).

Atoms are mostly empty space. Even though it has nearly all the mass, the nucleus is incredibly tiny. If the helium atom shown in Figure 4-4 were a mile across, the nucleus would be only about the size of a golf ball. If all of the atomic nuclei in the Earth were combined into one ball, it would be only about 200 yards across. The reason for this discussion is to point out just how close the two protons in the helium nucleus are. Being of like (sign) charge, and being so close, the protons repel each other mightily. In fact, this repulsive force is about 100 pounds! That is a huge force for such a tiny particle (10 trillion protons laid side by side would stretch only 1 centimeter.) One might wonder what holds these protons in the nucleus. The answer is that there exists an extremely powerful force called the "nuclear force," or it is sometimes called the "strong force," or even the "strong nuclear force," that can hold protons and/or neutrons together. It *is* an extremely strong force, but it can reach out only a tiny fraction of the diameter of an atom.

Notice that if a proton and an electron could be combined, their electric charges would cancel out. This combination can happen; the resulting particle is a neutron. Now notice that four hydrogen atoms have all the ingredients to form one helium atom. When this combination occurs some mass disappears. The mass of the helium atom is about 0.75% less than the combined mass of the four hydrogen atoms. By $E = mc^2$ this small mass loss has been converted into energy which must have left the scene. Without the energy leaving the scene there would have been no apparent mass reduction! This last point is very important, and it will be further discussed later.

The matter in the universe is about 75% hydrogen (by mass, even greater by atom count). Almost all the rest is helium. Normal stars have about this mix of elements, so they have quite a supply of hydrogen fuel. The energy production of our Sun reads like a recipe:

> 600 million tons of hydrogen   (only one ingredient)
> Bake at 15 million degrees C   (extremely hot oven)
> Yield: 595.5 million tons of helium plus 4.5 million tons of energy

How much is 600 million tons? To give some idea, it is about twice the weight of the human population. It is also about the weight of 1/7 cubic mile of water. (Take a section of land - one square mile, and cover it with water to a depth of 750 feet.) Since humans have about the same density as water (we barely float), we could take all the people in the world and stack them like cordwood on one square mile of land; the stack would be less than 400 feet high. Of course, everyone would want to be on the top layer.

You may have noticed that the recipe did not say how long to bake. Actually, the Sun constantly bakes many trillions of times this much hydrogen, but "only" about 600 million tons converts (reacts) to helium **PER SECOND!!** "What? You mean that the sun is using its fuel at the rate of 600 million tons *per second*? How much longer can it last?" Well, the sun has been doing this for about 6 **B**illion years, and in all that time it has used only about 5% of its supply of hydrogen. Stars are just a whole lot bigger than anyone can imagine. Bigness is essential for their functioning. The news is not all good, however. This current, quiet, peaceful stage of the Sun's life will last only about 5 billion more years. During this peaceful stage of their lives stars have access to only about the 10% of their hydrogen fuel that was originally in their cores. After that, the Sun will swell to a red-giant star which will possibly engulf Earth. It will burn us up whether or not Earth is engulfed.

Remember that 100 pounds of anything converted to energy would supply the United States for a year. 4.5 million tons per second gives the Sun an energy output that is truly "astronomical." (Pun intended.)

Why does it have to be so hot for this reaction to take place? From Section 4-3 recall that temperature is a measure of the *random* speeds of the particles that comprise an object. Temperature is actually proportional to the average kinetic energy of the particles. (Again: It should be realized that temperature does not *cause* these kinetic energies. The presence of random kinetic energies *is* what temperature *is*.) Because of the randomness of these kinetic energies and momentums caused by all of the collisions that occur, some very small fraction of the particles will (momentarily) possess many times that average kinetic energy. Applying this: At a few tens of thousands of degrees collisions will strip the electrons off the hydrogen atoms. With the electrons gone, the protons are now free to get together and react, but their electric charge causes a very large repulsive force (remember the 100 pounds of force between the protons). The 15 million degree temperature is required so that some protons on a collision course will have enough speed to overcome this repulsion and get close enough together for the nuclear force to grab them and slam them together. Falling together under this most powerful force, similar to water falling through a hydroelectric dam, is what actually releases the energy. In this more complete view of the hydrogen-to-helium reaction, so much energy is produced, and leaves, that the remaining mass is noticeably reduced. For the Sun, most of the energy that leaves is obviously in the form of light. If the energy did not leave the scene, then its mass equivalence would still be present, so there would be no apparent mass reduction. Energy weighs! This is the important point mentioned at the beginning of this section.

In our earlier discussion we did not mention how the energy leaves the microscopic scene when the protons fuse together in the core of the sun. We ignored the fact that at each step of the fusion process other particles are produced. These include positrons (anti-electrons), neutrinos, and very high energy photons called gamma rays. These low-, or zero-mass particles, rather than the more massive protons and neutrons, carry away the bulk of the energy. This is similar to shooting a gun. At the instant the bullet leaves the barrel, the gun has an equal momentum backward (the recoil or kick) to what the bullet has forward. This result is from conservation of momentum. However, the more massive gun has much less kinetic energy than does the small bullet, so the bullet can do more damage. [At low speeds, $KE = \frac{1}{2} mv^2 = p^2/2m$, so with the same momentum ($p = mv$); the smaller mass will have the greater kinetic energy (smaller number in the denominator).] This is also true at high speeds where we must use the relativistic expressions for both momentum and kinetic energy. Appendix A13-2 shows that this result is true for all speeds. In addition the positrons will not go far before they combine with an electron and completely annihilate each other, turning their combined mass completely into energy in the form of two additional gamma rays. This last process is an example of a matter-antimatter reaction that drives some spaceships in science fiction. It *is* the most efficient energy production method available. No other process can *completely* convert matter into energy.

The combination of high temperature and nuclear reactions which combine the nuclei of atoms gives us the term "thermonuclear fusion." One normally hears this term associated with H-bombs. In a bomb, one cannot wait for billions of years for the reaction to occur. The reaction must be completed in less than a microsecond. This is accomplished by not using a paltry 15 million degrees, but rather about 75 million degrees. This temperature is provided by an "ordinary" uranium bomb. The hydrogen-to-helium reaction in stars is usually done in three steps. The first step is by far the slowest. To really speed things up in an H-bomb we use deuterium (heavy hydrogen - where the hydrogen nucleus has one proton and one neutron) and tritium (still hydrogen, but with one proton and two neutrons in the nucleus). Because only part of the fusion reaction is performed in our H-bombs we get only about half the energy from the hydrogen in our bombs that the complete reaction gives to stars. But even so, we still get to make really powerful bombs.

4-8 The Perfect - Electric - Vehicle Question

Suppose that we have an electric vehicle. It will produce no emissions of any chemicals, nor any heat emissions. If it starts from rest relative to the road, and it runs off internal charged batteries, then will it weigh more after it has accelerated to a speed of $c/2$ (or any high speed) than it weighed when at rest? Consider this question from the rest frame of the road. Make sure that you think about this question as thoroughly as you can, then consult Appendix 18 for the answer and additional information.

In Special Relativity, we cannot consider a situation from a frame that is accelerating. An observer in the car will certainly feel acceleration, so at the level of this book we will not be able to properly analyze the situation from the accelerating rest frame of the vehicle. However, we will be able to answer some questions from the viewpoint of an accelerating observer by using the results from the next chapter.

4-9 Summary

This chapter ends our development of Special Relativity that we have derived from our first two postulates. We have found that certain concepts and laws that were thought to be inviolate prior to relativity now need modification. Time and space become relative. Momentum, an extremely useful concept when one considers many situations, needs to be redefined by dividing the classical definition by the relativity factor, R. Much more shocking was the discovery that mass and energy are really two forms of the same entity, now called "mass-energy." We derived the famous equation, $E = mc^2$, and saw that it has two meanings depending on whether m is the rest mass or the relativistic mass. Even our definition of kinetic energy, the energy of motion, takes on a different form from the classical result. Special Relativity has also provided us with other new ways of viewing the universe. All of this has sprung from the seemingly innocent Postulates 1 and 2 which make two simple statements about motion.

Chapter 5 – Gravity

As it happens, gravity and relativity seem to be connected. Also, we have seen numerous times that relativity has a strong connection with light. To adequately describe gravity, we first need to cover two more topics concerning light. These topics are: "colors" and the "Doppler effect."

5-1 Colors

In Section 1-3 we learned that when light travels it has the characteristics of waves. However, unlike sound or water waves, nothing material wiggles; instead light consists of oscillating electric and magnetic fields. The electric and magnetic fields are perpendicular to each other and also perpendicular to the direction of travel of the light. For example, if a laser beam is shining upward and its electric field oscillates north and south, then its magnetic field

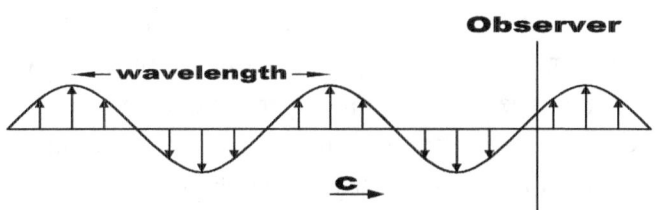

Figure 5-1 A drawing of a light wave moving to the right. This is shown in the rest frame of the observer. The wavelength of a wave is the distance from one point in the wave to the next similar point.

will oscillate east and west. By convention, when we sketch a light wave it is the electric wave that is shown. Figure 5-1 shows an oscillatory wave that is moving to the *right* at speed *c*. At the instant shown, the observer will be detecting a small upward, but decreasing, electric field. If you have trouble seeing it this way, think of the wave as stationary and the observer (the vertical line) moving *left*. Remember, only *relative* motion has meaning.

The *frequency* of a wave is the number of oscillations (complete wiggles – up and down) that pass by an observer each second. Light that is visible to human eyes (i.e., *visible light*) has a frequency in the range of about 200 to 400 trillion cycles per second. Radio waves, like light, are also electromagnetic waves, but their frequencies are much lower, only millions of cycles per second. The frequency of the microwaves in your microwave oven is about 2.5 billion cycles per second. (2.5 GHz, see Appendix A1-5 for an explanation of "GHz". A "Hertz", or Hz, is a cycle per second.)

Primate, bird, and day-flying insect eyes (and some other animal eyes) can roughly sort photons according to their frequency. The frequency of a single photon is proportional to its energy, and it is the energy of the photon

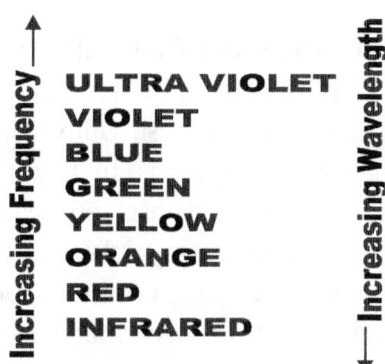

Figure 5-2 The visible (to human eyes) spectrum plus UV and IR with the types (colors) of light and their frequencies and wavelengths.

that eyes actually measure. This energy information is passed to the brain, which produces sensations (colors) corresponding to the different photon energies (different frequencies). We have **named the sensations. These are the familiar spectral colors and are shown in Figure 5-2. There is nothing red about red light, nor is there anything blue about blue light. These are only impressions supplied by the brain**. In the order of increasing frequency the visible colors are: red, orange, yellow, green, blue and violet. Violet light is almost twice the frequency of red light. The colors of visible light are similar to musical notes and cover nearly one octave (a doubling in frequency). Light of higher frequency than violet does exist. It is now much better known to the public than in the past. It is called *ultraviolet light*, or simply UV. It causes suntanning, sunburning, skin cancer, and the fading of paint and clothes. About 98% of the UV from the sun is blocked by the ozone layer in Earth's upper atmosphere, a percentage that has been decreasing in recent decades. Besides energy, photons also carry momentum. See Equation A13-8. Each of these is proportional to the frequency of the photon. Thus an ultraviolet photon carries both more energy and more momentum than any visible photon. *Visible* photons do not *individually* carry enough energy to damage living tissue, but a UV photon does. That photons are emitted and absorbed individually is what Einstein explained to win his Nobel Prize. (Not for relativity.) The sometimes wave-like and other times particle-like nature of light is called "wave-particle duality."

At the other end of the visible spectrum lies *infrared radiation*. It carries energy and momentum like all forms of *electromagnetic radiation*, but we cannot see it. We can feel it as the warmth from a fireplace.

Notice the term *wavelength* in the previous figures. This is simply the distance from one wave crest to the next, or from one trough to the next, as shown in Figure 5-1. The higher the wave frequency, the shorter the wavelength, and vice versa. (See Figure 5-1 and the first two paragraphs in Appendix A14-1.) This is because in vacuum all colors of light move at the same speed, *c*.

5-2 The Doppler Effect   (also known as Doppler Shift)

If an observer is moving toward a light source, she will pass more wave crests per second than if she were at rest. Conversely, if she and the light source are separating, she will *be passed* (light always outruns us) by fewer wave crests per second than if she were at rest relative to the light source. This shift in observed frequency (and also wavelength) is called the *Doppler Effect*.

You might say, "Wait a Minute! I thought that we said that the speed of light was not affected by the motion of its source. What gives?" That is correct. The *speed* is not affected, but the *frequency* and *wavelength* are affected by motion (as are also momentum and kinetic energy of the light). Since only relative motion has any meaning (Postulate 1), then it does not matter which (source or receiver) is considered to be moving. The Doppler Shift in sound is more familiar to us. The sound of an approaching vehicle is heard at a higher pitch than we hear after the vehicle has passed and is receding. The Doppler Effect for sound is actually more complex than for light. With sound it does matter slightly whether it is the source or the observer (listener) that is moving because sound moves at its characteristic speed relative to the air through which it is traveling, rather than relative to the listener (the observer), as light does.

Important: For the remainder of this section, the descriptions will all be in the rest frame of the "receiver" of the wave or signal. The formula for the relativistic Doppler Effect for light is:

$$frequency_{received} = \frac{frequency_{atrest}\sqrt{1-v^2/c^2}}{1+v/c} = frequency_{atrest}\sqrt{\frac{c-v}{c+v}} \qquad (5\text{-}1)$$

where the speed, v, is *positive* if the source and receiver are moving apart. (receding) [See Appendix A14-1 for a derivation of this relativistic Doppler shift formula.] The middle expression in Equation 5-1 is the more versatile. If the light emitter is *not* moving straight away from or straight toward the observer, then v in the numerator (in the relativity factor) must be the total speed, while v in the denominator must be the component of the velocity straight away from (+) or straight toward the observer (−). Appendix A4-4 explains "components." If the receiver is moving away from the source (positive v), a lower frequency will be received (called a *red shift* because red light has the lowest frequency of any visible light). Motion toward the source will produce a *blue shift*. (It would have been more correctly called a "violet shift.") Relativity contributes only the relativity factor, R, in the numerator. This is simply time dilation (slowing of time); the object that produces the waves will wiggle slower as determined by the moving observer (receiver or viewer). *At low speeds* where relativistic effects are very small the final square root in Equation 5-1simply approaches $1- v / c$ or $1/(1 + v/c)$, the latter being the form that Christian Doppler (1803 - 53) derived for sound, but in that case c (in the two forms immediately above) will be the speed of sound, and when dealing with sound the Relativity factor is practically equal to the value one, so we may replace it with the value "one" in the middle form of the Doppler equation for sound. The wavelength of light is actually expressed more often than its frequency. The wavelength is usually denoted by the Greek letter (lower-case) lambda, λ. The wavelength and the frequency are related by $\lambda = c / f$. If we take Equation 5-1, invert all three expressions around the equal signs, then multiply all three expressions by c, then we obtain Equation 5-2 below. See Appendix A14-1 for some help on this. The relativistic Doppler effect in terms of wavelength is:

$$\lambda_{received} = \frac{\lambda_{at\ rest}(1+v/c)}{\sqrt{1-v^2/c^2}} = \lambda_{at\ rest}\sqrt{\frac{c+v}{c-v}} \qquad (5\text{-}2)$$

With a positive velocity (receding), v, the numerator in the final fraction, is greater than the denominator, so the fraction and the square root are each greater than one. This means that for the source and the observer receding from one another, the received wavelength will be longer. We may think of this as the wavelength of the wave getting stretched out by the increasing separation between the source and the receiver. Appendix A14-1 has the derivations of the relativistic Doppler formulas.

For the case of the source and receiver approaching each other (negative velocity), the wavelength will be compressed. To obtain this conclusion from Equation 5-2: In the final square root, the numerator will be less than the denominator, hence the square root will be less than the value one, so the wavelength will be smaller (more compressed) than that seen by an observer at rest in the rest frame of the emitter.

For more information on the use of Equation 5-2, refer to the discussion immediately following Equation 5-1, but replace "higher frequency" with "shorter wavelength", and replace "lower frequency" with "longer wavelength". Table 5-1 shows some examples of the Doppler effect on signals that travel at c. This includes visible light and all forms of radio waves.

Table 5-1 The Doppler effect on light-speed signals as calculated for Frequencies (Eqn. 51) and Wavelengths (Eqn. 5-2).

v is the relative speed between the source and receiver.

v	$f_{received}$	$\lambda_{received}$
0	f_{rest}	λ_{rest}
Source and receiver separating (receding):		
$+0.1\,c$	$0.905\,f_{rest}$	$1.105\,\lambda_{rest}$
$+0.5\,c$	$0.577\,f_{rest}$	$1.732\,\lambda_{rest}$
$+3c/5$	$1/2\,f_{rest}$	$2\,\lambda_{rest}$
$+0.9\,c$	$0.229\,f_{rest}$	$4.359\,\lambda_{rest}$
$+0.99\,c$	$0.071\,f_{rest}$	$14.107\,\lambda_{rest}$
$+0.999\,c$	$0.022\,f_{rest}$	$44.710\,\lambda_{rest}$
$+c$	0	∞
Source and receiver approaching:		
$-0.1\,c$	$1.105\,f_{rest}$	$0.905\,\lambda_{rest}$
$-0.5\,c$	$1.732\,f_{rest}$	$0.577\,\lambda_{rest}$
$-3c/5$	$2\,f_{rest}$	$1/2\,\lambda_{rest}$
$-0.9\,c$	$4.359\,f_{rest}$	$0.229\,\lambda_{rest}$
$-0.99\,c$	$14.107\,f_{rest}$	$0.071\,\lambda_{rest}$
$-0.999\,c$	$44.710\,f_{rest}$	$0.022\,\lambda_{rest}$
$-c$	∞	0

From the results shown in Table 5-1, notice that any amount of change in either the frequency or the wavelength of a signal can be achieved by the observer or source moving at an allowed speed, that is, moving at a speed less than c. Once more, only *relative* speed has any significance.

A Doppler-shift example: A man was ticketed for running a red traffic light. He had some knowledge of the Doppler effect and tried to outsmart the judge by explaining that the traffic light may, in fact, have been *red*, but because he was approaching the red light its color was Doppler shifted to a higher frequency, or shorter wavelength, which happened to be *green*. So naturally he proceeded through the intersection. Unfortunately for him, the judge had read this book and knew the Doppler formulas, Equations 5-1 and 5-2, and after some pencil and paper calculations said to the defendant, "OK. I will drop the charge of running the red light, but I will have to fine you for speeding. At the very low rate of $1 per mile per hour over the speed limit, your fine is 200 million dollars." (Two hundred million miles per hour! This is approximately 30% the speed of light! No wonder we don't notice the Doppler effect on light the way that we do on sound. We can drive over 10% of the speed of sound, but nowhere near that fraction of the speed of light, and it is this fraction, v/c, that enters into the calculation as can be seen in the intermediate expression in Equations 5-1 and 5-2.)

Police radar measures your car speed using the Doppler effect. Also, most of us have heard of the Doppler weather radar. The microwave pulse from the radar reflects off the solid (snow, hail or car) or liquid (water) and a small fraction of the pulse returns to the radar unit. The equation for the Doppler shift upon *reflecting* off a moving object would be almost the same as Equations 5-1 and 5-2, but would require that we square the multipliers of frequency and wavelength. This is because the Doppler effect happens once when the signal arrives at the target, then again (multiplied on top of itself) when the signal is re-emitted by the target. For the low speeds encountered in most earthly events, reflection off a moving object is very close to replacing v with $2v$ in the Doppler formulas, Equation 5-1 and 5-2. This may be the calculation that is done in some earthly Doppler radars and is sufficient until speeds reach about 15% of c. Bringing up this very fine distinction between doubling and squaring to a traffic judge is not recommended. For there to have been a 1% difference between these two calculation methods for the indication on a police radar, your speed would have to have been about half that of the driver above. That would be admitting to a speed that would result in a 100 million dollar fine. The important point is that an object's speed toward or away from the radar can be measured from the frequency shift of the echo. A tornado is indicated by rapidly approaching objects (hailstones, houses, Dorothy), close to rapidly receding objects (raindrops, Toto, witches). We will make use of the Doppler effect in the next section.

5-3 The Equivalence Principle and Gravity

In our discussion of relativity so far there seems to be no hint that anything is wrong. In fact, an extreme amount of confirming data has amassed over the past century. However, some of what we are about to study may be on shaky ground. We need more and better experiments to help us find the true nature of the universe.

We are now prepared to start our study of the phenomenon of gravity. Gravity is a weighty subject, but we will consider it only in an elementary way. Almost all of our treatment of gravity can be done by working with an effect that we have all experienced: When an elevator begins to move upward, the riders feel heavier. The same increase in apparent weight occurs when an elevator that is moving downward stops at a floor. Oppositely, when the upward-moving elevator stops at a floor, the passengers feel lighter briefly, as they do when an elevator starts moving downward. People and animals are not bothered much by the feeling of increased weight, but the feeling of decreased weight is frightening to us because this is the feeling we have when falling – an instinctive fear.

We will summarize the ideas in the previous paragraph with our third and last postulate which is called the "equivalence principle." It is the familiar g-forces idea that gravity and acceleration feel the same, but with two slight embellishments:

Postulate 3: No experiment can distinguish between uniform acceleration and uniform gravity.

The word "uniform" is used before "acceleration" to mean that the ride is smooth; the rocket engines produce no vibration, and the acceleration is not changing magnitude or direction. "Acceleration" is the rate of change of velocity (miles/hour per second) and is a vector quantity (Section 4-1)

The units of acceleration are usually written as a length divided by time-squared. For

Figure 5-3 Two Labs are aboard rocket ships. The left ship is far from any planet or star and is firing its engine to accelerate upward (in the figure) with an acceleration, a. The right ship is standing on the surface of a planet whose gravity will produce a downward acceleration of g on unsupported objects. These two laboratories will obtain identical results for all experiments if $a = g$.

example, Earth's gravity is 9.8 meters / sec^2 (same as 9.8 meters per second per second). It is easier to understand these units if we do not express the speed and the time using the same time units. For example, suppose that we are driving and increase our speed (accelerate) by five miles per hour each second, or 5 miles/hour per second, or 5 miles / (hour · second) (read "five miles per hour per second"). Since an hour is 3600 seconds, we can express this acceleration as 5×3600 miles / hour^2 or 18,000 miles / hour^2. This sounds more impressive but is the same acceleration.

"Uniform gravity" restricts us to fairly small experiments, not in a spaceship that is hundreds of miles wide or long. On a big enough scale one could detect that Earth's gravity is pointing in slightly different directions (always toward the center of Earth), or that it is weaker higher up. The two labs shown in Figure 5-3 should obtain identical results for all experiments if $a = g$ (both considered as positive). That is, if the left ship with its engine firing is accelerating *upward* at the same number of meters per sec ^{2} as the ship at right would accelerate *downward* if it were *not* supported by the ground. We *must leave the support from the ground in place* for these two situations to be indistinguishable.

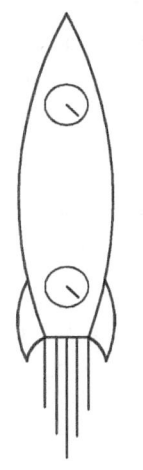

Figure 5-4 Accelerating spaceship with clocks in nose and tail

The fact that gravity and acceleration feel the same is one of the effects used in virtual-reality rides. For example, a ride may tilt your seat back gradually, and move the scene on the screen up to mask the tilting. Since you are leaning back gravity presses your back into the seat back. Then when you see acceleration on the screen you really feel like you are accelerating. To further enhance the effect, they may even blow air in your face, even though there would be no wind inside a closed, accelerating vehicle.

For our study we need to realize that the equivalence principle states that experiments done inside a spaceship far from any planet or star but accelerating at 1g (the strength of Earth's gravity at its surface) will give identical results as that same ship at rest on the launch pad on Earth. The acceleration caused by Earth's gravity is g = 9.8 meters/sec ^{2} or 9.8 meters per second per second. This means that a dropped object will increase its downward velocity by 9.8 meters per second each second that it falls. For experiments done on the surface of our Moon, the spacecraft would need to accelerate at only about 1.6 m/sec ^{2} to give identical results. Our moon's gravity is weaker than Earth's gravity because Earth is more massive.

Consider the spaceship shown in Figure 5-4. It is very far from any star or planet that might cause a gravitational attraction. Unlike the spaceships drawn in earlier chapters, this ship is not coasting at a constant velocity; it is accelerating upward. That is, its upward speed is increasing. [The sculpted vertical lines at the bottom are not "zipples" (the cartoon lines used to show motion), but rather the rocket engine exhaust, indicating acceleration upward. This will be the case for all rockets for the remainder of this chapter.] This ship is equipped with two highly accurate atomic clocks; each clock has light emitting and detecting equipment. The lower atomic clock causes a light signal of frequency 300 trillion cycles per second (green – the frequency being controlled by the lower atomic clock) to be beamed toward the upper clock and associated light detecting apparatus. The signal is emitted near the tail, but by the time the light signal reaches the upper clock the whole ship and clocks have accelerated to a higher speed than when the light signal was emitted. The upper end of the ship now has a higher upward speed than the source did when the light was emitted at the tail. This will produce a Doppler shift in the frequency of the received light. It will be shifted to a lower frequency (That is, a red shift; the frequency is measured using the upper atomic clock). Another way of thinking of this is: To the light signal, the separation between the two clocks has effectively increased because while the light was in transit the upper clock moved upward compared to where the lower clock was when the light signal was emitted. A third way to think about this red shift is that the increased separation between the point of emission and the point of reception has effectively stretched the space between the clocks, thereby stretching the wavelength of the light while it was in transit.

By Postulate 3, the Equivalence Principle, this Doppler shift, caused by the acceleration of the ship, is equivalent to the effect of gravity. Thus, when light moves upward through gravity it will be red-shifted. The energy and momentum that a photon carries are both proportional to its frequency (Section 5-1). Thus, as photons climb out of a gravitational field they will lose both energy and momentum, but they will *not* slow down; their frequencies will simply decrease and their wavelengths will increase. We can look at this in another way. A photon carries energy and this gives the photon a mass equivalence. It is very familiar that a mass thrown upward loses kinetic energy as it rises in a gravitational field.

Similarly, but oppositely, a light signal emitted by the nose clock will be received by the tail clock at a *higher* frequency than when it was emitted because by the time the signal reaches the tail, the tail will have accelerated to a higher upward speed compared to that of the light source when the light was emitted. A *blue shift* will be observed by a tail observer. In Section 2-3 we saw that if we have a head-on collision with a photon it will not be moving any faster to us. However, we now see that the photon will be Doppler shifted to a higher frequency and shorter wavelength, hence shifted to a higher momentum and energy. After all, this is like falling in a gravitational field, but no photon can speed up or slow down; its speed is always c (in a vacuum).

We can calculate how much shift in frequency or wavelength there will be. For the next two paragraphs we will be in the (accelerating) frame of the spaceship. Call the vertical separation between the clocks D (in the rest frame of the ship). Light will then require a time $t = D / c$ to travel this separation. By the definition of acceleration: Acceleration is change in velocity / time interval; so a change in velocity = acceleration × time interval. (We multiplied each side of the first equation by "time interval" and turned the equation around. If the ship's upward acceleration is a, then the velocity of the upper clock will have increased by $\Delta v = at = aD / c$. Inserting this relative speed between the upper and lower clocks (receding, therefore positive) into the Doppler formula, Equation 5-1, we obtain:

$$frequency_{received\ at\ nose} = frequency_{emitted\ at\ tail} \sqrt{\frac{c - aD/c}{c + aD/c}} = frequency_{emitted\ at\ tail} \sqrt{\frac{1 - aD/c^2}{1 + aD/c^2}} \qquad (5\text{-}3)$$

In the last step, we have divided both numerator and denominator of the fraction by c.

The frequency of the emitted light will be controlled by the adjoining atomic clock, and similarly for the detected light frequency. Thus, an observer by the *nose* clock, who was not thinking relativistically, expected to receive a frequency of 300 trillion cycles per second. He will conclude that the tail clock is running *slow*. Similarly, an observer by the *tail* clock will conclude that the nose clock is running *fast*. This is *not* like the case in Section 1-9 where each observer is moving at a constant speed and each determined that the clocks in the other reference frame ran slow. We are in a single, but accelerating, frame here. Since time is that which is measured by clocks, then time itself must be running faster in the nose of the ship than at its tail, as long as the engine is firing or the ship is resting on the launch pad in a gravitational field. Again, using the equivalence principle, we conclude that time, and clocks, will run faster on the upper floors of a building than on the lower floors. The ratio of these clock rates (time rates) is given by the square roots in Equation 5-3 with the acceleration due to gravity (9.8 meters/second^2) inserted for the acceleration, a. D is the vertical separation between the floors in question. [*For small values* of a and/or D, the square root in Equation 5-3 is more easily calculated by the approximation $(1 - aD/c^2)$. This is similar to the approximations in Appendix 11.] Note that the previous two paragraphs were considering the tail (or lower in gravity) clock and its time as they are determined by an observer beside the nose (or higher in gravity) clock . For observations of the nose clock, and its time, by an observer beside the tail clock, simply interchange all the signs in this and the previous paragraph $(+ \Leftrightarrow -)$.

People have fun with this connection between time and elevation:

Greetings! Dear brother from Boulder,
I hear that you've gotten much older.
And please tell me why
My lower left thigh
Hasn't aged quite as much as my shoulder.

(David Morin, www.physics.harvard.edu/limericks.htm , used by permission)

In 70 years, only 0.0004 seconds more time would have passed in Boulder (one mile up) than at sea level; Earth's gravity is weak, so the effect is small. 70 years x gD/c^2 = 0.0004 seconds, where g is the acceleration due to Earth's gravity (9.8 m/s^2); D is one mile = 1609 meters; and 70 years = 2,209,000,000 seconds. At higher elevations one's speed due to Earth's rotation is somewhat faster. This higher speed would cause time to slow down as determined by an observer at sea level. However, this effect is much smaller than that caused, in the opposite direction, by Earth's gravity.

For this paragraph we will not be in any particular frame. If all of this section so far seems a little inconsistent with some previous results, then refer to Section 1-5. We called the relativity that we were developing at that point "Special Relativity." It is *special* (limited in some sense) as opposed to *general* relativity (applicable in all cases). The property that is *special* about what we studied before we looked into gravity is that our reference frames, from which we are *determining*, or *making measurements*, must NOT be accelerating. (Note that accelerations are felt as g-forces.) We may consider accelerations of things within a frame, but the *observer* must not be accelerating. Another way of saying this is that in special relativity something at rest in a reference frame will remain at rest. This is clearly not the case here on Earth. If we let go of something it will fall because of gravity [which is equivalent to acceleration (Postulate 3)]. However, gravity on Earth is extremely weak compared to the electrical and magnetic forces that manipulate the atomic particles that we use every day. An example is the electrons that are shot toward the screens of old computer CRTs and TV sets. In such cases, gravitational effects are so tiny compared to the electromagnetic forces that gravity can simply be ignored, and the reference frame here on Earth seems to be essentially not accelerating (by Postulate 3, essentially not in gravity for these really fast particles).

Figure 5-5 A light beam crossing an accelerating spaceship

For the remainder of this section we will be back in the frame of the accelerating spaceship. Now another effect: Consider the accelerating spaceship shown in Figure 5-5. It is far from any massive body which might cause a gravitational attraction. A light beam is fired straight across the ship. It does take some time for the light signal to cross the ship. During this time the ship will have increased its vertical speed because it is accelerating upward. The light will be slightly left behind by the accelerating ship and will hit the right wall a little lower than straight across the ship. Therefore, using the equivalence principle, light is deflected downward by gravity. How far downward will the light beam be deflected? If the ship is w wide it will take light a time of $t = w/c$ to cross the ship. If the ship is accelerating at 1 g (Earth's gravity at its surface), then from high-school physics, an object starting from rest will fall a distance of $\frac{1}{2} g t^2$ in a time t. Inserting the time for light to cross the ship we obtain $\frac{1}{2} g w^2 / c^2$ as the downward deflection of the light. If this ship, or room on Earth's surface, is 10 meters wide, this deflection will be about 5 trillionths of a millimeter. This is comparable to the diameter of a typical atomic nucleus.

Postulate 3 claims that the ship's acceleration is equivalent to gravity. This amount of bending has not been directly measured on Earth. However, astronomers have photographed light being deflected as it passes astronomical objects. For example, starlight bends in passing near the sun, and light from very distant galaxies is bent upon passing near a foreground galaxy. These effects appear on photographs of galaxies looking something like eyebrows, and are sometimes called "Einstein Arcs." The effect is named "gravitational lensing."

5-4 Resolution of the Twin Paradox

We will now use some of what we have discovered about gravity to resolve the famous Twin Paradox. The Twin Paradox from Section 3-7 arose when (before we had enough knowledge) we looked at the situation in Roamer's (the traveling twin's) frame. Without a knowledge of gravity we reasoned incorrectly that in Roamer's frame it was his Earth-bound twin, Homer, that should have been the younger. From Roamer's point of view it was Earth that moved away and returned. Roamer seemed to be at rest in his own frame, except for his brief accelerations. It was mentioned in Section 3-7 that the solution to this paradox had to wait until we had learned about gravity.

Look at Figure 5-6. From Earth to A Roamer is accelerating away from Earth. From A to B he is coasting at constant speed. From B to C Roamer is accelerating toward Earth. By the

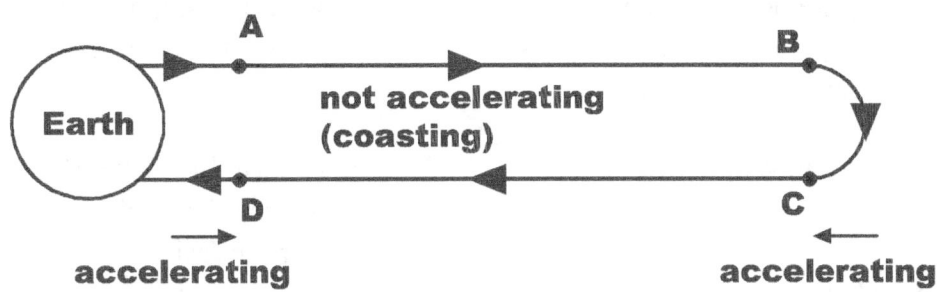

Figure 5-6 Roamer blasts off Earth (accelerates to the right), then coasts for a long time (in either frame), turns around (accelerates to the left for twice as long as the blast off), coasts back toward Earth for as long (in either frame) as he coasted on the outward journey, then accelerates to the right again – braking his speed on arrival back on Earth.

Equivalence Principle (Postulate 3) this is as if Earth, and his twin, are very high in a gravitational field. Time passes more rapidly high up, and Earth is light years up, so the time on Earth passes very rapidly to the traveling twin. How all this will *appear* to each twin will be covered in Section 6-8. Here we are *determining* the situation (god's-eye view). This is like looking at Figure 5-6 from far enough above the page so that all the events are essentially equidistant from us, thus we have equal signal-time delays from all points in the experiment. At point C the traveling twin, Roamer, would actually determine (accounting for the changing signal-travel times) that there was an even greater age *difference* between himself and his twin than when he finally returns to Earth. Homer, the Earth-bound twin, is the older. Again coasting from C to D Roamer seems to be at rest while his Earth-bound twin is rushing toward him at very great speed. Thus, from C to D, Homer, the stay-at-home twin, ages less than Roamer and Rover (in the frame of the traveling twin, remember). Finally, from D to Earth, the traveling twin again accelerates away from Earth (stopping). In both Earth to A, and from D to Earth, the traveling twin is accelerating away from Earth. By the Equivalence Principle this is as if Earth were low in a gravitational field. Clocks should run slower on Earth, but unlike the turnaround far from Earth, here there is very little vertical distance (D in Equation 5-3) between the two twins, so this portion of the trip contributes very little to the time differences. When these effects are all taken into account, the traveling twin will arrive at the same conclusion as his stay-at-home twin: The stay-at-home twin will be older than the traveling twin and by the same amount as calculated much more easily by Earth-bounders. In Chapter 6 we will examine what each twin actually *sees and experiences* of the other during various portions of the trip, that is, if they do not take into account the effects of the signal travel times.

These twins are actually quite famous. In some treatments of relativity they are given names such as "Peter" and "Paul." This is okay, but the reader tends to forget which twin is doing the traveling. This is why we used the names, "Homer" and "Roamer". (Houser, his trusted companion and Rover may, or may not be twin dogs.) Besides being instructive, the twins have inspired their own Limericks. Here are two examples:

I've heard of the traveling twin.
It causes my head to spin.
As to which is older,
I'll say even bolder,
"This is the weirdest that's been."

Sadly, Twin Brother in pine,
From space I'm writing this line
To your son Jack.
And when I'm back,
I hope I'm still older than mine.

Using today's technology, would this method of traveling into the future be a practical one? It is easier to consider the question in Homer's frame. Let us assume that we could actually launch a spaceship from Earth orbit at a speed of 1% c (that is, $c / 100$). This is 1860 miles per second and is about 100 times the speed of any spacecraft that has ever been launched by humans. The financial and energy requirements for this trip would strain the global budgets. If Roamer spent 50 years total on the trip, 25 years each way, on his return he would find that Homer was slightly less than one day older than himself. Why so little difference? At this speed, the relativity factor, R, would be 0.99995 according to Table 1-1. $1/R$ is 1.00005. This latter number is the rate that Roamer would be traveling into the future compared to Earth time. (1.00000 is the rate that Homer is traveling through time.) The rate of aging difference between the twins will be 1.00005 - 1.00000 = 0.00005. If we multiply this rate difference times 50 years, we get 0.9 days. Roamer would spend 50 years in the ship, the resources of Earth would be significantly reduced, and Roamer would have traveled into the future less than one day. It would be much more effective for Roamer simply to take a day-long nap. If he did take the trip, he probably would not get to live that extra day because during 50 years in space he would have been exposed to quite a high dose of cosmic radiation. We will leave it to the reader to decide the merits of an actual "Traveling-Twin Project."

5-5 General Relativity – Curved Space-Time

Einstein thought that the loss of energy and momentum when light climbs out of a gravitational field, and especially the downward deflection of light as it moves across a gravitational field, might be results of curved space caused by a massive object such as Earth. To make the idea work, time must be considered as a fourth dimension. To put time on an equal footing with the three space coordinates, time must be multiplied by a constant that has the dimensions of meters per second. This is the dimension of speed. What constant speed do you think might be a good guess?

By now you would probably guess, *c*. This turns out to work. Once you have the general idea, this is not really "rocket science."

In making this curved (warped) space-time concept work, it turns out that all objects must move at *c* in space-time. An object at rest in space (in a reference frame) is doing all of its 4-D motion along the time axis. If this object were moving through space, then to keep its total speed at *c*, it must not move at *c* along the time axis, but at a lesser speed (along the time axis). This is exactly the slowing of time that we got in Special Relativity (see Appendix 15). Light already moves at *c* in space. This means that a photon has no motion along the time axis at all. In order for General Relativity to be truly *general*, it must reduce to special relativity in the absence of nearby massive objects, i.e., no gravity.

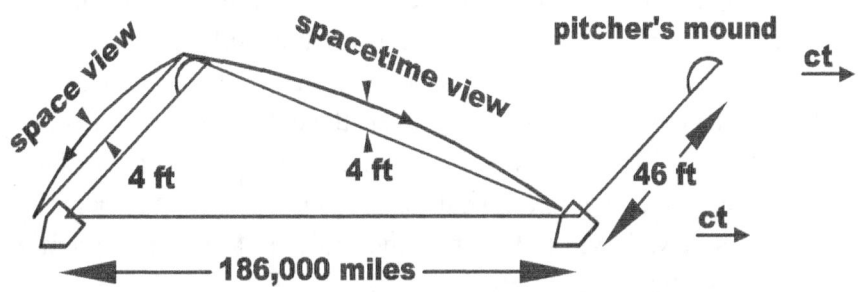

Figure 5-7 A slow-pitch softball game in space-time. Time is to the right.

We cannot draw 4-D figures, or even visualize them. However in some simple situations, we may use one of our space axes to plot the time axis, actually *ct*. Consider Figure 5-7 which is a game of slow-pitch softball. Suppose that the pitched ball takes one second to travel from the pitcher's mound to home plate. During one second the ball will rise 4 feet, then fall 4 feet. We will now use an axis to the right as the time axis. During this one second the whole Earth and ball will have moved 186,000 miles to the right on the time axis. The space-time path of the pitched ball will not be the simple four-foot rise and fall and the 46 ft. of horizontal motion, it will also move 186,000 miles down the time axis. The amount of curvature of space-time caused by Earth's *mass* is this four-foot deviation along a slightly diagonal path length of 186,000 miles. This is an extremely slight curvature, but then Earth's mass is very small in the astronomical scheme of things.

You may hear or read that the curvature of space-time is *caused by gravity*. This is not correct. **In Einstein's theory, gravity is an effect *caused by curved space-time*; the curvature is produced by the proximity of *mass*, or its equivalent, *energy*.**

Figure 5-8 shows another example of curved space-time: Earth orbits the Sun once per year. The diameter of Earth's orbit is 186 million miles. Light would require about 1000 seconds (16.7 minutes) to cross the diameter of Earth's orbit. This orbit is Earth's path in *space*. During a year the entire solar system moves very nearly a light year (six trillion miles) along the time axis. If we use the space direction perpendicular to Earth's orbit to plot the time axis, then Earth's path through space-time is a very stretched out slinky (a helix). This almost straight path is the curvature of space-time that the mass of the Sun causes at Earth's distance from the Sun. One turn of the helix would actually be about 100 thousand times the diameter of Earth's orbit along the time axis.

Yet another "186" coincidence: Earth's orbital speed, 18.6 miles per second, is almost exactly one ten thousandth of the speed of light. That is, c is 10,000 times the orbital speed of Earth. These same coincidences would also work in the metric system, but they would be "3"coincidences. (c = 300 thousand km/sec, and Earth's orbit is 300 million km in diameter.)

According to General Relativity (GR), objects that have no forces acting on them move along these paths that are curved by the presence of masses. These paths are called *geodesics*. In GR, gravity is NOT a force. Near a massive object these geodesics are curved toward the massive object. If someone is about to jump from the roof of a building, the building is exerting an upward force on the person. During the time that this upward force from the building is acting on the hesitating jumper, the path of this person in space-time is **NOT** a geodesic. After leaping, the path of the person through space-time IS now a geodesic. However, the sidewalk below is being pushed upward by the soil and rock below so it is NOT on a geodesic. The curved paths of the person and the sidewalk intersect a couple of seconds later. That is, the sidewalk hits the person. Splat! The paths through space-time of the sidewalk and the roof of the building are along essentially parallel paths that do not intersect unless the building collapses.

In the case of the slow-pitch softball game, during the time the softball is in flight, it is considered to travel a geodesic, free of any forces acting upon it. Rather than thinking that it is acted upon by a gravity force, it merely travels a path through space-time that is curved by the presence of the earth's mass.

5-6 Event Horizons and Black Holes

In the space-time gravity theories (There are variations on Einstein's) the question arises about combining lengths. Basically, the question is "How much space plus how much time equals how much space-time?" The answer is usually expressed in an equation called a "metric." These metrics usually look a lot like the Pythagorean theorem (Appendix A4-1 and A4-2) that we used in Section 1-9 and explained in Appendix 15. There we used the Pythagorean theorem in two and three dimensions, but in flat (uncurved) space-time the Pythagorean theorem is true in any number of dimensions. In multi-dimensional space the theorem is like before, except that "The square of the hypotenuse is equal to the sum of the squares of *all of* the other sides." In 4-D space-time, there would be four other sides.

In flat (no curvature, no gravity) space-time, the metric would be (see Appendix 15):

$$ds^2 = dx^2 + dy^2 + dz^2 - c^2 dt^2 \qquad (5\text{-}4)$$

where *ds* is the infinitesimal distance in space-time, *dx* stands for an infinitesimal change in the *x* component of position, *dy* and *dz* are similar, and *dt* is the infinitesimal time interval that passed during the change in space position. Remember, time must be multiplied by *c* to give it the same character as the space dimensions. (Or multiply time by *ic* where *i* is the square root of −1. See Appendix A3-2 for a brief discussion of imaginary numbers and Appendix 15 for a discussion of the minus sign in Equations 5-4 and 5-5.) Appendix 15 also shows that Equation 1-2, the slowing (dilation) of time, our quintessential relativity result, is derivable from (is contained within) Equation 5-4. Referring to Section 3-3 on causality, the right side of Equation 5-4 can be seen to be $D^2 - c^2 t^2$, where the sum of the three space terms (each squared) is a distance, *D*, in space (squared), and the time term is a "distance in" or "length of" time (squared). Remember, this causality form, called the "metric" here, has the same value in any reference frame – a very useful property.

When space-time is curved, causing the phenomenon of gravity, the terms on the right side of Equation 5-4 must be multiplied by factors. The square roots of these factors are essentially the degree that time slows down near a massive object (similar to Equation 5-3), and another new effect, the distortion of space near a massive object. In Einstein's GR, the metric becomes (the mathematics here is well beyond the level of this book. The time term is usually written first):

$$ds^2 = -\left(\frac{1-\Phi/2}{1+\Phi/2}\right)^2 c^2\, dt^2 + \left(1 + \Phi/2\right)^4 \left(dx^2 + dy^2 + dz^2\right) \qquad (5\text{-}5)$$

where:

$$\Phi = \frac{G\,M}{c^2\, r} \quad , \quad \Phi_{\text{Sun surface}} = 0.000002 \quad , \quad \Phi_{\text{Earth surface}} = 0.0000000007$$

Φ is the "gravitational potential" caused by a spherical mass (star, planet), *G* is the *gravitational constant* (from Isaac Newton days), and *r* is the distance from the center of the spherical mass, *M*, to the point where the metric is applied. (*M* is the mass *inside* a sphere of radius *r*.) The values of Φ on the spheres' surfaces are easy to calculate from a knowledge of the mass and radius of each object. (The gravitational constant is $G = 6.67 \times 10^{-11}$ meters^3 / kg sec^2, the mass of the sun = 1.99×10^{30} kg, Earth's mass is 5.97×10^{24} kg, Sun's radius = 6.96×10^8 meters, and Earth's radius = 6.38×10^6 meters.) These very small values for Φ are the reason that we do not easily notice the effects of curved space-time in the solar system except for the obvious attractive nature of gravity. There may be other, more subtle effects of gravity than the readily apparent one.

Notice that the squared multiplicative ( )^2 factor times the dt^2 term in Equation 5-5 can become zero if the mass is large enough and/or the distance *r* is small enough so that Φ = 2 (In this case the numerator becomes 1 - 2/2 = 0.) This could happen if the mass of our sun were squeezed into a sphere only a few miles across, so that r can become small and yet remain outside the massive sphere. (Look back at what M is.) This multiplying factor in the time term going to zero is the prediction of the famous *event horizon* (black hole). With zero multiplied times the time-interval term, dt^2, it would appear that an infinite amount of time could pass with no *change*, dr, in the distance, r, from the center of the compact object. That is, a particle falling in would simply appear to hang at this distance forever. Also, any point-size particle that comes so close to such a compact body such that $\Phi = 2$ would never be able to escape (according to Einstein's GR). Of course, nothing is point size so the situation is more complicated. Quantum mechanics seems to provide a mechanism for allowing particles to both fall in and to very gradually leak out. Any star that begins life with more than about eight times the sun's mass is expected to eventually collapse when its hydrogen fuel is exhausted to the size where Φ can equal 2. (and form a black hole?)

It is simple to calculate the radius of the event horizon predicted by GR if we know the mass of the compacted star, and also know that the radius of the compacted star is smaller than the calculated radius of the event horizon: We simply take the definition of the gravitational potential, Φ, from above, set it equal to 2, divide each side by 2 and multiply each side by r. We obtain:

$$ r = \frac{GM}{2c^2} = 0.74 \ km \times \left(mass \ in \ solar \ masses \right) $$

For example, if M equals 5 solar masses, then the radius of the event horizon is predicted to be 3.7 km, or 2.3 miles. For Earth to become a black hole, it would have to be compressed to the size of a marble (about 1 cm in diameter). It seems unlikely in the extreme that this could ever happen to Earth, but for a large enough star, its gravity alone might provide the needed compression. We could say that this gives the reader an idea of the compression required, squeezing five times the mass of the sun into a ball less than five miles in diameter. It is difficult to comprehend this magnitude of compression, but it should be noted that neutron stars have more than one solar mass in a diameter of about twelve miles.

Before finishing his gravity theory, Einstein had expected that the factors that multiply dt^2 and dx^2 in the metric (Equation 5-5) would be the exponential functions $e^{-2\Phi}$ and $e^{2\Phi}$ respectively. [See Appendix 5 for a description of the exponential function. Some competing gravity theories do in fact have these exponential functions in their metrics. In Appendix 5, it is shown that the factors in the GR metric match these exponential functions to first order; i.e.,in terms raised to the first power (Φ^1, but not Φ^2 nor at higher powers of Φ) . See the latter part of Appendix 5.] With the small values of Φ encountered in the solar system, we cannot *experimentally* determine which multiplicative ( ) factors are correct. If the exponential functions are correct, the multiplier of the time term will never go to zero, meaning no event horizons and no black holes. So, the prediction of black holes, or not, depends on the ( ) factors that multiply the dt^2 term, and these factors depend on the mass of the central object and on which theory of gravity is being considered. The exponential function and the GR time term are compared in Table 5-2.

In the case of zero mass in the vicinity, Φ will be zero, and the agreement is exact. This is special relativity. The difference between the two terms does not become important until about $\Phi = 1$. At $\Phi=2$, the GR term becomes zero, so an event horizon is predicted by GR, but not by the theories of gravity which have the exponential functions in their metrics. The exponential factor never goes to zero, no matter how large the value of Φ. The largest value of Φ to be found in the solar system is that near the sun, with $\Phi = 0.000002$. Here the factors are so close to each other that experimental results have essentially no chance to give any gravity theory a good test on this basis.

Table 5-2 Values of the exponential and the GR time ( ) functions for various values of Φ :

Φ	$e^{-2\Phi}$	$[(1-\Phi/2)/(1+\Phi/2)]^2$	
0	1	1	(zero gravity, exact match)
0.000002	0.999996	0.999992	(solar surface, indistinguishable)
0.1	0.81873	0.81859	(still really close)
0.21	0.65705	0.65603	(neutron star surface)
0.5	0.36788	0.36000	
1.0	0.13534	0.11111	(observationally distinguishable ?)
1.5	0.04979	0.02041	
2.0	0.01832	0	[black hole (0), or not]

5-7 Is General Relativity Correct?

Einstein did not believe that GR was completely correct. However, what he, or any other person *believes*, is of no importance. The indication of correctness is if the theory agrees with all experiments.

Einstein could think of only three feasible tests for GR. (1) Gravitational red (or blue) shift of light. We have already discussed this in Section 5-2. It *is* observed, and GR predicts the effect to within experimental accuracy. (2) Bending of starlight as it passes near the sun. Again, this has been observed during solar eclipses and agrees with experiments to within experimental accuracy (which is not very high). And, (3) a very small shift in the orbit of the planet Mercury is predicted and observed, but only after several rather large corrections from Newtonian mechanics are applied. These corrections should be OK, leaving a small and reasonably well known effect.

Since the time of Einstein, others have used his theory to predict a very small (3 cm ≈ 1 in.) oscillation in the Moon's *apparent* orbit around Earth that has been detected by bouncing laser beams from Earth off reflectors left on the lunar surface by astronauts. Some competing theories of gravity died because they failed to predict this small apparent oscillation. This all sounds fairly good for GR, but other competing theories of gravity have appeared which also reproduce all of these results to within experimental accuracy. Sometimes these other theories will predict very slightly different answers, but still within experimental accuracy. As a result, we have no decisive proof of the superiority for GR, but it is widely believed to be correct.

However, there *are* difficulties with GR. Remember, a theory is not necessarily true just because many people believe it to be so. GR seems to be incompatible with two other physical theories. These are Quantum Mechanics and some aspects of Electromagnetism. The combination of Quantum Mechanics and Electromagnetism was accomplished about 1950. This theory is called Quantum Electrodynamics (QED). QED has predicted some effects which agree with experiment to an astounding 12 decimal places. To be incompatible with such stunning successes indicates that there is a problem somewhere. There are some other theories of gravity which may be compatible with QED and which also pass all of the previously mentioned gravitational tests. You may ask, "Why don't we replace GR with one of these newer paradigms?" One reason may be that GR had no serious rival theories for so long, about 70 years, that it is extremely entrenched. Such a large number of researchers have used it for their whole careers that they are understandably reluctant to give up GR without decisive proof of its inadequacy.

GR predicts that a black hole cannot produce a magnetic field outside its event horizon. However, observations of binary star systems where one of the members is thought by most investigators to be a black hole do show evidence of very strong magnetic fields. Much astrophysical work is currently being done in this area in attempts to resolve such issues.

5-8 Toward the Future: Gravity Probe-B and Other Gravity Theories

The Gravity Probe-B (GP-B) experiment was under planning, development, construction and testing for over 40 years. It is a satellite experiment that was launched on April 20, 2004 into an earth orbit that passes directly over Earth's poles. This satellite is probably the most precisely constructed device ever conceived by humankind. Ground-based resources such as widely spaced radio telescopes are also involved. The experiment is expected to measure, with extreme accuracy, very subtle effects of gravity called the "geodetic effect," which can be interpreted as a curvature of space-time (or in other ways in some other theories of gravity) which is caused by the mass of Earth, and "frame dragging," caused by the spin of Earth on its axis. Some of the various gravity theories do predict different values for these two effects, even zero for frame dragging, so a precision experiment could very well direct us to the correct theory, if it exists yet. Hopefully the GP-B experiment will be highly successful because there is no other area in physical science that is in such dire need of precise experimental results to guide our thinking. The results of the experiment should be available about April 2007.

[**Note added at press time (2014)**: The GP-B experiment has returned data of lesser quality than hoped, but good enough to confirm the geodetic effect predicted by GR to about one percent and the frame-dragging result to about twenty percent. The experiment has a very good web site: einstein.stanford.edu (no www). The next paragraph describes the situation as assessed by Prof. Jones as of late 2007. SLR]

The GP-B results could range from ho-hum to revolutionary. If the results agree with GR, then some people will unwisely wonder why we did the experiment at all. However, if the results disagree with GR, especially the frame-dragging result, then we might usher in a new era in our understanding of one of the most important aspects of nature, namely gravity.

To put the question of the correctness of GR into a historical context, consider what happened after the theory of electromagnetism was discovered. This theory was essentially completed in about 1865 by James Clerk Maxwell after very considerable work earlier by others. It was found to be incompatible with Newton's mechanics, which was developed a couple of hundred years earlier. As far as could be told at that time, Newtonian mechanics seemed to be a perfect description of nature. Some physicists wondered what was wrong with (the new) electromagnetism. It predicted a correct value for the speed of light, but without specifying any particular reference frame. We know now that any frame will do. Decades later Einstein took a different approach. He wondered what might be (very slightly) wrong with Newtonian mechanics, and special relativity was born. At that time no one realized that the Theory of Electromagnetism was already relativistically correct. *That* was its incompatibility with Newtonian mechanics. In Section 1-5 it was stated that our development here was not going to follow the historical sequence in which relativity was actually discovered, so we have mentioned history only slightly. The lesson to be learned from this bit of history is that no matter how well entrenched a theory (or idea) has become, we should always be willing to consider new thinking that might supplant older ideas.

An example of an alternate gravity theory that is quite different from GR, that agrees with all experimental results and seems to be compatible with both Quantum Mechanics and Electromagnetism has recently been developed by K. Krogh of the University of California, Santa Barbara. Instead of curved space-time, Krogh's theory postulates that the speed of light, c, is reduced near massive bodies. Krogh's theory uses the same gravitational potential, Φ, as GR. Einstein tried this idea before he developed GR. One major developmental difference between GR and Krogh's theory is that Quantum mechanics was not developed until about a decade after GR, so Einstein's thinking could not benefit from this discovery. Quantum Mechanics extends the wave-particle duality of photons (Sections 5-1 and 1-11) to all objects. Its effects primarily show up in the world of the very small (protons, electrons and atoms), but then these make up larger objects. (Sections 5-6 and 6-7 have more about Quantum Mechanics.)

In addition, Krogh has postulated that quantum mechanical waves are also slowed and their wavelengths are reduced near massive bodies. This essentially doubles some effects. Without these Quantum Mechanical effects, the calculation for the deflection of starlight by the sun (one of Einstein's suggested tests) would give only half the measured deflection. The mathematics of Krogh's theory is much like that of Optics and Electromagnetism. (The theory of Optics is derivable from Electromagnetism.) While Krogh's theory is not what is called a "metric theory," an equivalent metric derived from his theory has the exponential functions that Einstein expected. [Equations 5-4 and 5-5 are the metric equations for *special relativity* (SR) and *general relativity* (GR), respectively.] With the exponential functions in his metric, Krogh's theory does *not* predict the existence of Black Holes, nor does it require a Big Bang to have occurred. Since it is not a curved-space-time theory, Krogh's theory predicts a zero frame dragging result for the GP-B experiment.

The other measurement to be made by GP-B is the "geodetic effect." In GR this effect is a consequence of curved space-time. Although Krogh's theory predicts essentially the same value that GR does, the underlying mechanism is different. The math level for these predictions is beyond that appropriate for this book. There are other competing gravity theories, but this brief description of one such theory indicates some of the types of changes to our thinking and understanding that the results of the GP-B experiment, and possibly other experiments, might bring about. If we continue to experiment and develop theories, then the passage of time should eventually reveal and/or confirm the truth about gravity, one of the principal "forces" in the universe.

[**Note added at press time (2014)**: As a result of the Gravity Probe B measurement of "frame-dragging", the original version of Krogh's theory must be abandoned; however, it might be modified to encompass the result by incorporating "gravitomagnetic" effects in the theory. Such effects are similar to those of magnetic fields that are caused by the motions of electric charges. The motions of masses produce analogous gravitational force effects. Their inclusion is necessary for even Newtonian gravity to be made compatible with special relativity. Prof. Jones included the discussion of an alternative gravity theory because he believed that it might eventually be necessary to modify general relativity in order to produce compatibility between relativistic gravity theory and quantum mechanics. Revisions are both normal and sometimes necessary for science to progress. SLR]

Chapter 6 - Visual Appearance and Time Travel

This chapter is mostly for fun. It is primarily descriptive and contains the fewest equations. One new consideration will dominate the topics covered here, that being the time for a light-speed signal to travel from different parts of the "experiment" to the observer. This is not really a relativistic effect, but some already-learned relativity results will still be needed, and additional relativity results will be developed.

6-1 The Visual Appearance of a Distant Rapidly Moving Object under Uniform Illumination

The question that we will consider here is, "Can we actually *see* length contraction of an object that is moving at some significant fraction of the speed of light?" If we were in space where some object such as a house moved past us at half the speed of light, we would certainly not see it at all; our senses and brain functions are far too slow. But, what if we used a hyper-camera with an extremely fast shutter to take a snapshot of this passing object? The *still* camera image is formed from the light that passes through the shutter during the very brief time that it is open. Thus, the image that the camera will produce will *not* be corrected for the time differences taken for light to arrive at its lens from different parts of the passing object that are at different distances from the shutter. Our eyes also form images from the light that arrives at an instant. This process is what we mean by the word "see" as opposed to the word "determine" that we have used previously. The distinction between these two terms will become clearer as we investigate further.

The following example is slightly tedious to follow, but the rewards for doing so are great. Take heart! We will consider Figure 6-1 which shows a block (a cube) that is 5 meters (5 m) on an edge as measured in the rest frame of the cube. This choice for the size of the cube will simplify our discussion. The block will move past us at a speed of $v = 4/5\ c$. The face that is leading (in front) will have an "A" on it. The side that will face us when it is closest to us will have a "B", the side opposite B will have an "F" (far side), and the trailing side opposite "A" will have a "C." The side on top will have a "T." The only relativistic effect that will concern us here is length contraction (not, for example, the slowing of the rotting of the block, nor its increased mass, momentum, or kinetic energy). Referring to Table 1-1, we see that for this speed the relativity factor, R, will have the value of 3/5, so the block will be contracted (face A to face C) to a length of 3 m (= 5m × 3/5). Side B will still be 5m tall, but will be only 3m long (right to left in the figure, that is, in the direction of motion).

The *non-relativistic* effect that we *will have to apply* is what was mentioned above, the differences in the transit times for light from different distances to reach the camera. For example, it will take longer for light from side F to reach the camera than that from side B. Thus, **we will see where side F was at an earlier time [5 meters divided by c (= 16.7 nanoseconds) earlier] than where we see the nearer side B.** Actually, the camera will not be able to see side F since this is a solid, non-transparent block. If the cube were a wire frame then we would see this effect. It is an important consideration, however, because the far edge of side C is at the same distance from the hyper-camera as Side F. Surprisingly, we *will* be able to see side C, even though it is not facing the hyper-camera. We shall define t_{5m} to be this time of 16.7 nanoseconds required for light to travel 5 m, the edge lengths of the cube at rest. We will need this value in several instances.We will begin by considering a simplified situation that assumes that the camera is sufficiently far from the block and using a telephoto lens, so that the light that leaves the block headed toward the hyper-camera will be traveling along essentially parallel lines. (This over-idealization is commonly used in books and articles, and it will be further analyzed later.) For additional simplicity, we will consider only one position of the block, namely its closest point to the camera where the light from the block that is headed toward the camera will be traveling in a direction perpendicular to the direction of the block's motion

In Figure 6-1 the upper third is two top views of side T of the contracted block. This could be either

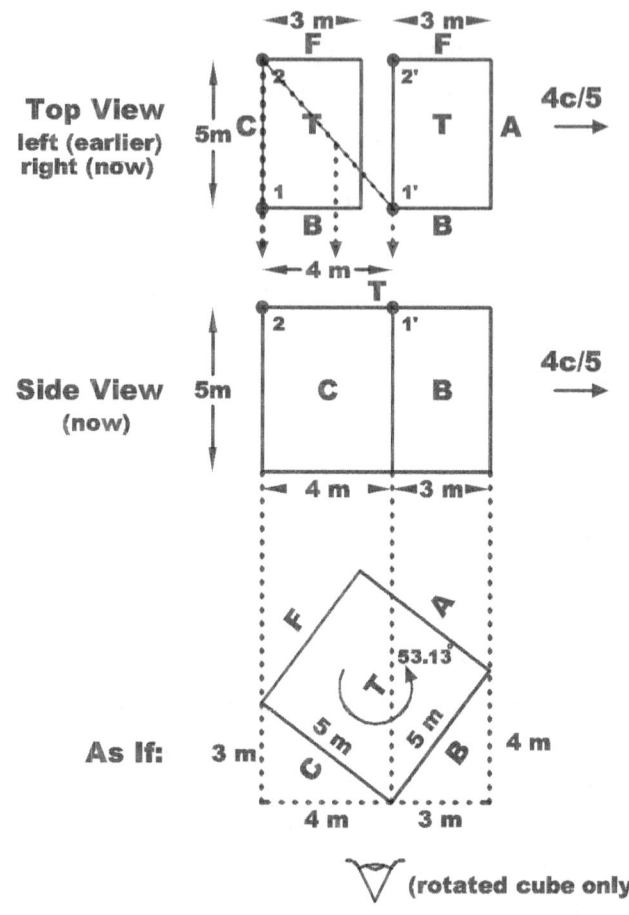

Figure 6-1 Constructions to answer the questions about the visual appearance of a cube that very rapidly passes by us.

"determine" or "see" because all points on side T are equidistant from the observer in this view. This view is *not* the view that we are analyzing for the visual appearance; it is a construction to help in understanding the situation. Our hyper-camera is looking straight at side B. For the time being, completely ignore the bottom, rotated square and the eye. This is not a view, but it will be used later as an analogy.

(The following four paragraphs should be read slowly and carefully while referring to Figure 6-1.) Still considering the top view in Figure 6-1: The Lorentz contraction of sides B, T, and F, in the direction of motion, to 3 meters is shown. Vertical edge 1 (the edge between sides B and C) is a dot in this view. Light that starts from edge 2 (the edge between sides C and F) that happens to travel straight toward the camera will get to the camera because face C is moving to the right and gets out of *this light's* way. Repeating an important point, it will take light that is leaving edge 2 a time of t_{5m}

to reach the plane of face B. During this t_{5m} time, the contracted block will have moved to the right a distance of 4 meters [$4c/5$ (the block's speed) multiplied times t_{5m} equals 4 meters]. This position is shown in the right half of the top view. At this position and time, light leaving from edge 1 (now shown as 1') that is moving toward the hyper-camera will be even (head to head, or neck and neck) with the light that departed edge 2 a time t_{5m} earlier, so they will reach the hyper-camera at the same time. The instant that these two light signals reach the hyper-camera the shutter clicks. This produces the image shown in the "side" view of Figure 6-1. Light from other places on side C that is heading toward the hyper-camera will be spread right to left between the edges 2 and 1': The diagonal beaded line from edge 2 to edge 1' is a hypothetical vertical plane as seen from above. Light that leaves from side C that is moving toward the hyper-camera, and happens to leave at the time that side C intersects this plane, will also be even (head-to-head) with light that left edge 2 and is *now* even with the other arrowhead points. ("*now*" = the time since the light left from edge 2 and reached the arrow points.) Thus, the hyper-camera will be able to see side C (spread out, but slightly contracted right-to-left). Side B is at the right, shown Lorentz contracted to its length in the direction of motion, 3 meters.

Any light that leaves from side A cannot reach the hyper-camera because the motion of the block will cause side A to run into that light. This would either absorb or reflect the light. In either case the light from side A cannot reach the hyper-camera. Light from side A could reach an observer located mostly forward of the moving block, because this light will outrun the block.

The top view in Figure 6-1 shows only face T. As described immediately above, side C should be seen spread to the left in these views, but it is not shown. This top view is for construction purposes only.

Now consider the rotated square in the lower portion of Figure 6-1. It is a top view of a hypothetical cube with sides of 5 meters that has been rotated ccw through an angle of $53.13° = \sin^{-1} 4/5 = \sin^{-1} v/c$. (See Appendix A4-1.) The eye at the bottom shows the *direction* of view that the hyper-camera would have of this hypothetical rotated block. Notice that the vertical dashed lines match the above figures exactly. This indicates that the visual appearance of the moving block is exactly as if the block were rotated with the leading side turned away, but no *contraction appears* to have occurred. However, the side facing the camera, side B, is contracted by the expected amount. The unexpected appearance of side C gives the impression of rotation. If the hyper-camera were nearer the block, the rotated appearance would not be so convincing since the right side of face B is not seen as vertically shortened as it would if a real 3-D object were rotated. (In an actual rotation, the right side of face B would move farther from the hyper-camera or eye, and thus appear shorter. This will be further developed in Section 6-3.) Also, the left side of side C, edge 2, would appear vertically shortened for the same reason.

In case you wonder how nature could produce this illusion of rotation, go back to Section 1-9 on the transverse light clock. Notice that our relativity factor was derived by considering a right triangle. In the bottom portion of Figure 6-1, notice the 3, 4, 5 right triangles (see Appendix A4-1). In describing nature, the same math appears again and again.

Has this effect ever been seen? Even though the Moon is a sphere rather than a cube, this appearance of rotation will cause it to appear very slightly rotated as it orbits Earth. However, because the Moon's orbital speed is only 2000 miles per hour, the apparent angle of rotation will be only 0.0002°. Even if the effect were much larger, it would be very difficult to detect because this angle stays essentially constant, so all lunar features constantly appear at slightly incorrect lunar longitudes; we have never detected it. However, material orbiting near a neutron star would be moving extremely fast. This effect might show up in such a case, but since we only detect neutron stars as a point of light and x-rays, we really need to be very sure of our theory of gravity before we could definitely claim anything. There would also have to be some structure in the material in orbit that was very well understood in order to give us some effects to look for. More about this in the last paragraph of Section 6-6.

6-2 A Rapidly Moving Distant Object with Illumination from the Camera Only

What will the rapidly moving block look like if instead of uniform illumination from all sides, we take a photo with the light source positioned at the camera? The light source could be either a flash or a constant light source. There will be no eyes to give us red eye, but there will be considerations of which sides will receive light from the source and from which sides the reflected light can return to the camera. Side C will not be illuminated because Side B will leave a shadow going back left, and side C is moving into this darkness. Side A will run into light after this light has passed side B on the right and thus side A will be illuminated. However, light reflected from side A cannot return to the camera because it will be run into by the advancing side A (as before). Thus, what we will see is only the *contracted* side B, still 5 meters tall, and contracted to 3 meters wide, as shown in the right portion of the side view in Figure 6-1. Side C is *not* seen. This is probably what we would have expected to see before we thought to consider what light can reach the block and the camera.

6-3 A Rapidly Moving Object with Uniform Illumination as Seen with Depth Perception

Some of the material that we are about to cover might seem to be contradictory to what you may have seen, or will see, elsewhere. Some books and articles even claim that length contraction is invisible, but they are considering only the special case of the camera being very far from the path of the object, thereby eliminating depth perception. They also assume uniform illumination of the block. Except for the case of the light coming from the camera, this is what we have done so far in this chapter when we backed our hyper-camera off to a great distance so that it was receiving essentially parallel light from the block that passes by. We will now be bold and consider visual appearance *with* the assumption that the observer *does* possess depth perception.

You may have noticed that when using binoculars to view a person walking toward or away from you, they give the impression of marking time (walking in place). Optically, binoculars and telescopes magnify up-down and sideways more than front-to-back, just as the telephoto lens of our hyper-camera would do. The Moon, because it is so far away from us, in terms of the separation of our eyes, appears to be a flat disc rather than a sphere even without any magnification. Using only the parallel light from the object completely eliminates depth perception. The resulting flat appearance of the "rotated" object that was described in Section 6-1 is necessary to produce this illusion of "pure rotation."

Without considering depth perception, we got the appearance of a pure rotation when we looked at an object that was positioned at its closest point as it passed by at high speed. The analysis of the object at some general point on its path is too complex for this level, but it does yield similar results. The angle of apparent rotation increases as the object approaches and moves by. (If you are really interested, see Helliwell in the suggested Readings.)

We *can* perform a simple analysis of a 1-D stick that is moving straight toward us (or straight away from us) to show what happens *with depth perception*. We have mentioned that for observers to "determine" something from what they "see," they would have to correct their observations for the time of transit for the light signals. We will now show another example of what happens when we *do not* correct for the light transit times, nor will we take the god's-eye view used in most of the gedanken experiments. After all, neither our eyes (and brain) nor our hyper-camera does this correction. Again, this is what is meant by the word "see."

Consider a stick of rest length L moving away (left) from observer O at a speed v, as shown in Figure 6-2. A green light pulse leaves the far end of the stick and travels toward O. Let t be the

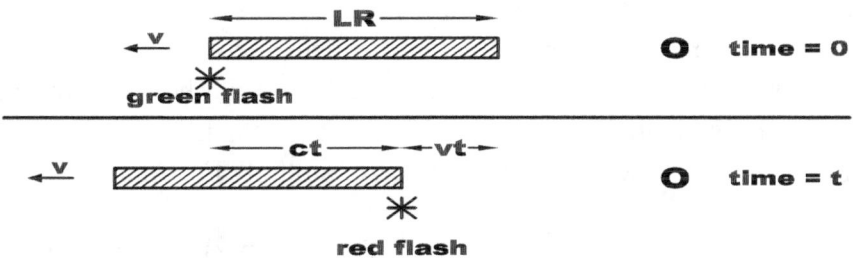

Figure 6-2 Measuring the length of a *receding* object: A stick of rest length L is receding from observer O. (upper half): A green flash occurs at the leading end of the stick. (lower half) When the green flash reaches the trailing end of the stick a red flash is emitted at time = t. The apparent length is the separation between the two flashes and is equal to ct.

time that the green pulse requires to reach the trailing (right) end of the moving stick which causes an immediate flash of red light at the right end. Both of these light pulses then travel at c toward O and will reach O <u>simultaneously</u>. This is important – this is what we mean by "see" or "appear." Observer O is equipped with a very precise rangefinder apparatus (a sophisticated apparatus similar to two widely spaced eyes, shown in Figure 6-3) so that the distance to the two flashes can be accurately measured. What apparent length will O *see* with her apparatus? We can calculate this by methods used several times before. This analysis is in the rest frame of observer O. Notice that the stick is contracted in length by the relativity factor R. The green light flash from the far end of the stick will travel a distance equal to ct in reaching the trailing end of the stick, where t is the time required.

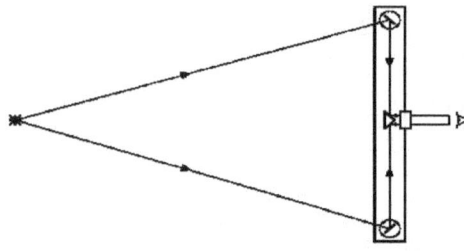

Figure 6-3 An optical range finder, as described in the text. The range (distance) is found from the angle of the mirrors and their separation.

During this time t the near (right) end of the stick will travel to the left a distance of vt (farther from the observer).

As can be seen in Figure 6-2, the contracted length of the stick, LR, must equal ct plus vt. That is:

$$LR = ct + vt \quad \Rightarrow \quad t = \frac{LR}{c + v} \qquad (6\text{-}1)$$

(see Appendices A2-2 and A2-3 on how to solve for t.)

The *apparent length* will be the separation between the green and red flashes since these two flashes occurred at the two ends of the stick, and the red and green flashes will arrive at the observer *simultaneously*. As can be seen in Figure 6-2, this apparent length will be less than the actual (contracted) length of the stick and will be simply ct, so we multiply Equation 6-1 by c:

$$L_{apparent} = ct = \frac{cLR}{c+v} = L\frac{c\sqrt{1-v^2/c^2}}{c+v} = L\sqrt{\frac{c-v}{c+v}} \qquad \begin{array}{l}(6\text{-}2) \; v \text{ is } + \text{ if receding} \\ \quad\quad\; v \text{ is } - \text{ if approaching}\end{array}$$

See Appendix A14-2 for the manipulations to arrive at the final expression in Equation 6-2.

The *apparent length*, ct, of the receding stick (positive v) is *less* than its length at rest. This can be seen in Figure 6-2 and is also indicated by the numerator in the final fraction in Equation 6-2 being less than the denominator, hence the fraction is less than 1, thus the square root is also less than 1. The shortening effect (on receding, i.e., positive v in Equation 6-2) caused by the unequal l signal travel times adds to the effect of relativistic length contraction; the receding stick will *appear* even shorter than its (actual) contracted length, LR.

Now consider the same stick *approaching* the observer as shown in Figure 6-4. Here we will temporarily call the speed of the stick positive even though it is approaching the observer.

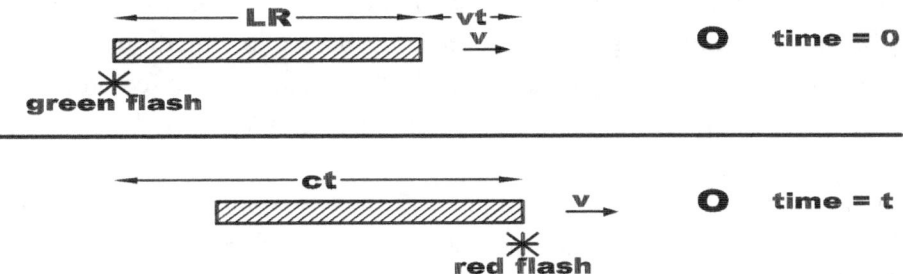

Figure 6-4 Measuring the apparent length of an *approaching* object: A stick of rest length L is approaching an observer, O. Upper figure: A green flash occurs at the trailing end of the stick. Lower figure: When the green flash reaches the leading end of the stick, a red flash is emitted at the near end of the stick at time $= t$. The apparent length of the stick to observer O is the separation between the two flashes, and from this figure, this is equal to ct.

As seen in Figure 6-4, we have the relation:

$$L\,R = ct - vt \qquad (6\text{-}3)$$

which is the same as Equation 6-1 for the receding stick, except that the sign on v is reversed. This demonstrates what usually happens when deriving a descriptive equation: We could have simply argued that all that was necessary for Equation 6-1 to handle the case of an *approaching* object is to follow our sign convention and use a negative sign on the approaching speed. Thus, Equation 6-2 will give the apparent length of either an approaching (negative v) or a receding (positive v) object.

An approaching stick, or the length of any approaching object, will still have its length in the direction of motion contracted by the factor R, but the light that the observer sees at some instant coming from the far end of the stick will have departed the far end at an earlier time than light from the near end. At this earlier time the approaching stick was farther away, so there is an apparent stretching effect on the *appearance* of the approaching stick. Now consider the math: Referring to Equation 6-2, on approach (negative v) the numerator becomes larger than the denominator, hence the square root is greater than 1. Thus, the approaching stick will appear lengthened; the lengthening effect of the signal travel times will outweigh the relativistic contraction of the stick's length at all speeds. (The use of a little math has saved us again when we have counteracting effects.)

On receding, both relativistic length contraction (Lorentz contraction) and the effect of the difference in light transit times have the effect of shortening the *apparent* length of the stick. (Remember, we are using the word "apparent" to mean "as it *appears* to the observer O, or the eye or camera.") On receding, the apparent length of the object is not only less than its rest length, it is less than its actual contracted (by R) length.

We can obtain an interesting result if we compare Equation 5-2 for the Doppler effect for wavelength lengthening of emitted light from a receding object to Equation 6-2 for the apparent shortening of a receding object. The lengthening and shortening factors are reciprocals of each other. ("Reciprocal" means 1 over the other, or "Turn the fraction upside down."). This means that if a yellow (at rest) car were approaching us at a very great speed, it would not only appear to be blue (Doppler shifted to a *shorter* wavelength, see Figure 5-2), but also *appear* to be a *longer* car, and by the reciprocal of the same factor. Then after the yellow (at rest) car passes by us and is receding, it might appear red (a *longer wavelength* of light), but appear to be *shorter.*

6-4 An Object Passing Near an Observer with Depth Perception and Color Vision

Let us reconsider the situation shown in Figure 6-1 of a cube under uniform white illumination moving past an observer. However, in the present case we will assume that the observer (or hyper-camera lens) is only 5 meters from the *center* of the cube at closest approach (2.5 meters from the near face of the block, side B). In this case the block will appear as shown in Figure 6-5.

Here, there are four effects to consider that did not affect our result when we viewed the cube from far back using only parallel light: 1) The far side (side F) will appear smaller in the photograph than the near side (side B). 2) The upper and lower edges of sides B and F will be farther from the

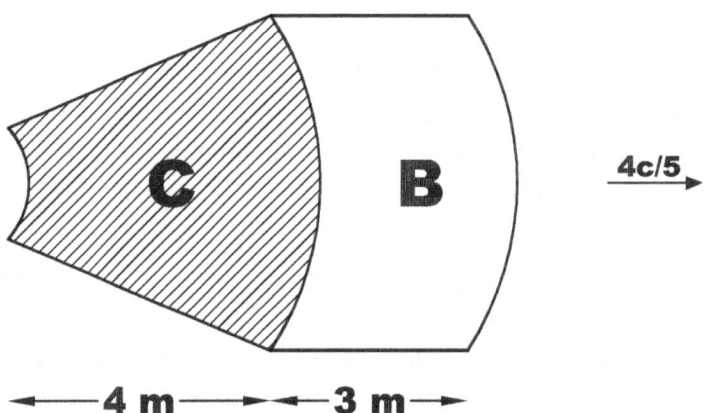

Figure 6-5 The 5-meter cube moving past a hyper-camera lens located only 2.5 meters from the center of side B. Uniform white illumination of the block is assumed. (This shape is derived from *calculations* of the signal delay times.)

lens than the centers of these sides. This means that light reaching the lens from the upper and lower edges will be from an earlier time than light from the center. Thus, the upper and lower edges will

112

appear swept back from the position of the mid-section of side B of the cube. This is also true for the far side (side F), but to a lesser extent than for side B because, being farther from the camera, there is less of a *difference* in the distances to the camera lens. 3) The trailing side of the cube, side C, will be illuminated only by the photons that happen to have velocity components to the right that are greater than the speed of the cube, 4c/5. Assuming completely random directions for the photons that are illuminating the cube, only 1/5 of the photons that would strike side C at rest will be able to catch up to side C. (This brightness calculation requires the use of integral calculus, so it will not be presented here, nor even in the appendices. This result is approximate because the motion of the block will leave fewer photons to the immediate left of the block, and this was not considered; side C will be even darker than calculated.) This means that side C will appear less than 20% as bright as side B. Side A would appear brighter than when it is at rest because it will run into more photons than would normally strike it when at rest, but this light cannot reach the camera, as explained in Section 6-1. 4) Light reflecting off of side C will be Doppler shifted toward longer wavelengths (red-shifted). Thus, in addition to the distortions in the apparent shape of side C, it will also appear to be a dim pink in color. This dimming and pinking of side C will also occur even when the cube is viewed from afar as described in Section 6-1, though this was not shown in Figure 6-1. (We are gradually including additional effects.)

So, the nearby passing cube shown in Figure 6-5 does still have a rotation component, but with the other effects included, it cannot be claimed to appear as a "pure rotation."

6-5 The Appearance of a Block Passing Near a Camera with Illumination Only from the Camera

We will begin by considering the situation where the camera lens and the assumed point-size light source is located 2.5 meters from the center of side B at its closest approach, that is, 5 meters from the block center. As explained in Section 6-2, light from this source can only reach sides A and B. As before, light reflected from side A will not be able to reach the camera lens because side A runs into any light from side A that reflects off it and is headed toward the camera. Thus, only side B will be visible in this case. When the camera is close we have the effects caused by the upper and lower edges, as well as the corners of side B being farther from the lens/light source than is the center of side B.

Figure 6-6 shows the appearance of the cube when the light source stays on, and the photo is taken by opening and closing a very quick shutter. Figure 6-7 shows the effect of a flash photograph. In a flash photograph, the shutter is held open for a while, but the very short duration of the flash "freezes" the action. In both cases considered here, there is a dimming of the illumination in the corners of side B because the corners are more distant from the light source than is the center of side B. Also, the light from the camera strikes the corners at a more glancing angle and spreads out more, further increasing the dimming effect. This dimming is shown as shading. In Figure 6-6, the swept-back look of the upper and lower edges is caused by light from these more distant edges taking longer to reach the lens than light from the center, so they are seen at their position at an earlier time

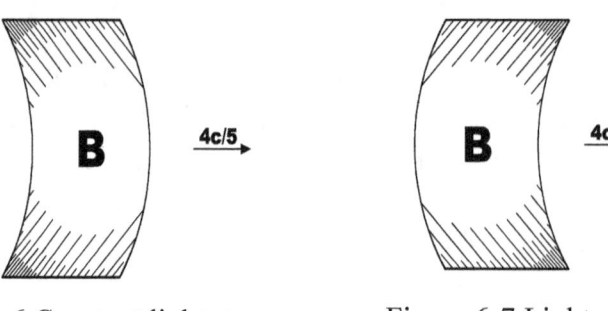

Figure 6-6 Constant light source located at the camera, quick shutter

Figure 6-7 Light source is a Flash located at the camera

than the center. This is as described earlier. A surprising result occurs when we take a *flash* photograph; see Figure 6-7. Light from the flash (located very near the lens) takes more time to reach the more distant corners of side B than is required to reach the center. The shutter is held open, so this reflected light does get back through the lens to the film. This has the effect of photographing the corners of side B at a *later* time than the center. This produces the swept-forward appearance of side B. At its near point, side B is neither approaching nor receding from the light source or film (or CCD if the hyper-camera is digital), so no Doppler shifts (color changes) appear.

6-6 Summary of the Appearance of Rapidly Moving Objects

With depth perception, an approaching object heading straight toward the observer will appear lengthened and blue shifted, while a receding object moving directly away will appear to be shortened and red shifted. Under uniform illumination an object passing by at its nearest distance from the observer will appear to be essentially rotated with the leading side partially turned away and the trailing side partially turned toward the observer; the trailing side will be dimmer and pink. In between these points there will be a mixing of the effects. Only rarely are the effects of depth perception, color shifts and non-uniform illumination considered when dealing with this subject. If the observer or camera is located near a passing object of any shape, then the distortions shown in Figures 6-5, 6-6, and 6-7 will also occur in their apparent image.

This chapter probably seems like it is just for fun, and that is much of the intent. However, there are some practical applications of what we have described. Giant accelerators produce sub atomic particles, usually protons, that move very near the speed of light. The particles are usually in clumps. It is both easier to accelerate a clump, and also some experiments need timing information that is available only when the clump of particles strikes a target. Control sensors that are off to the side of the motion will not detect the clump in the way that is simplest to use but will "see" the effects with the distortions and rotations described above. This is because the sensors detect electric and magnetic effects caused by the clumps of particles, and the electric and magnetic effects travel at *c*. Remember, light is *electromagnetic* radiation.

6-7 Apparent and Superluminal Speeds

"Apparent speed" simply means how fast a moving object *appears* to be moving to the eye or camera. "Superluminal speeds" means speeds that are faster than light. This is easy to say, but we have been contending that nothing can travel faster than *c*. What is meant by this term is motion that *appears* to be faster than *c*. Remember, in "appearance" we mean how the eye or camera would render some event(s) based on the light (speed) signals that arrive at the eye or camera at some instant without taking into account the different times of transit of the signals from parts of the event that are at different distances from the observer. When we did take the transit times into account, or used our god's-eye view as in most of the gedanken experiments, then we called the observation a "determination." We can use a result that we already have to arrive at an equation that will give us the apparent speed of a receding object.

Figure 6-8 shows a contracted stick of rest length *L* (as measured in its rest frame) moving away from the observer, O, at a speed, *v*. Now, all in the observer's rest frame: The stick will be contracted to a length of *LR*. A flashing device is located at a *fixed* distance from the observer. That is, it is at rest relative to the observer. It will emit a flash of light when either end of the stick passes.

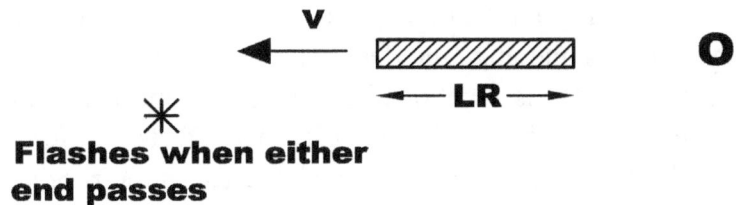

Figure 6-8 Measurement of apparent speed. A stick of rest length *L* is receding from the observer at speed *v*.

Since these two pulses of light must travel the same distance to reach the observer, the time interval between the arrival of the two pulses at O will be the time required for the stick to travel its length. As always, when considering relativity, to avoid possible mistakes, it is important that we carefully explain the details of the measurement process.

The *apparent* speed of the stick to Observer O will simply be the *apparent* length of the stick divided by the time it takes to travel its length (speed = length / time). From Equation 6-2 we have already found the apparent length of a receding object. Thus,

$$v_{apparent} = \frac{Apparent\ Length}{time\ to\ move\ its\ length} = \frac{L\sqrt{\dfrac{c-v}{c+v}}}{LR/v} \qquad (6\text{-}4)$$

One would expect that the apparent speed of the stick should *not* depend on the stick's rest length. This is verified since L divides out of the fraction. Appendix A14-3 shows that Equation 6-4 simplifies to:

$$v_{apparent} = \frac{cv}{c + v}$$

(6-5) v is positive when receding

Let us take a moment to examine Equation 6-4, the setup equation, and also the simplified result, Equation 6-5. As already mentioned, the rest length of the stick is immaterial to the calculation. We have already seen that the apparent length of an approaching stick will be greater than its rest length, while the apparent length of a receding stick will be less than its contracted (actual) length. Since the apparent speed is the apparent length of the stick divided by the time it takes to pass a point in space, then the effect of apparent length will *increase* the apparent speed of an approaching object, but will *decrease* the apparent speed of a receding object. The time required for the stick to travel its contracted (actual) length will be the same for both receding and approaching objects (with equal magnitudes of speed - because the contracted lengths are the same). In both cases, an increase in the magnitude of the speed will reduce the time required to traverse its length, so the faster the stick is moving, the faster the calculated speed. Notice that for an *approaching* object, both of the effects mentioned in this paragraph (apparent length and actual speed) *act to increase* the apparent speed. With these two effects reenforcing each other, the apparent speeds of approach are *greater* than their actual speeds. In the *receding* case, the reduced time to travel its length acts to *increase* the apparent speed, but the apparent shortening acts to *reduce* the apparent speed. Without the use of some mathematics, we could not determine which of these two counteracting effects is the larger. The math shows that the apparent shortening has the greater effect, so the apparent speeds of receding objects are less (smaller magnitude) than their actual speeds.

Table 6-1 shows some apparent speeds of approaching and receding objects as calculated using Equation 6-5. For speeds that we encounter in everyday life there is no noticeable effect from relativity. For approaching objects (negative v), the apparent speed is always greater (in magnitude) than the actual speed. For any approaching speed greater than half the speed of light, the *apparent* approaching speed will be superluminal (greater than c). The case of the photon is especially interesting. One way of interpreting the infinite apparent-speed result for an approaching photon is to realize that there is no way that you could be informed that a photon is coming toward you. Nothing, and no signal, can outrun the photon to warn of its arrival. Any light would seem to arrive instantly. There would be no way to dodge a pulse from a laser gun because by the time you could see the beam, or see the trigger pulled, the photons in the beam would already be to you! To contrast this situation with that of a sound pulse from a bomb, consider the situation of a bomb exploding in a wheat field. The observer could see the blast wave approaching at the speed of sound as it bends over the wheat. If the bomb were nuclear so its flash is bright enough to ignite the wheat, then the light from the burning wheat could not reach the observer before the light from the blast. This is similar to the scattered light from the ray gun beam.

Table 6-1 Apparent Speeds of objects moving straight toward or straight away from the observer as calculated from Equation 6-5		
v	$v_{apparent}$	
0	0	
approaching speeds:		
$-0.01c$	$-0.0101c$	(1860 and 1877 miles per second)
$-0.1c$	$-0.1111c$	
$-0.5c$	$-c$	(apparent light speed)
$-3c/5$	$-1.5c$	(superluminal)
$-0.9c$	$-9c$	(superluminal)
$-0.99c$	$-99c$	(superluminal)
$-c$	$-\infty$	(An approaching photon *appears* to have infinite speed!)
receding speeds:		
$0.01c$	$0.0099c$	
$0.1c$	$0.0909c$	
$0.5c$	$0.3333c$	
$3c/5$	$3c/8$	
$0.9c$	$0.473684c$	
$0.99c$	$0.497487c$	
c	$c/2$	(explained in the text)

For receding objects the apparent speed is always *less* than the actual speed. To understand the $c/2$ apparent speed for receding light, consider the following experiment: We fire a flashbulb and start our stopwatch. At a distance D from us is a small mirror that reflects some of the flash back to us. We stop the stopwatch when we see the returned pulse from the mirror. The stopwatch will read $2D/c$ since the light in the returned pulse has traveled a distance that is twice the distance to the mirror (there and back). However, "apparent" or "seeing" is *not* taking into account the travel times for the signal (the returning pulse that indicated that the light had reached the mirror), so we "miscalculate" the speed. (speed = distance over time = D over $2D/c = c/2$). But this *is* the speed that the light pulse would actually *appear* to be moving away from us. For example, if we had some smoke along the light path, then the pulse would *appear* to be moving away at $c/2$. This *apparent* speed could be deduced from the fact that the *apparent* diameter of a smoke-ring-shaped light pulse was decreasing at half the expected rate as it recedes through the smoke (assuming that we did *not* correct for the signal travel times).

Astronomers have observed superluminal apparent speeds when some violent event ejects a beam of matter (we will call it an "object") in the general direction toward Earth with a speed that is some large fraction of the speed of light, say, 0.894 c as in Figure 6-9. If the beam contains a clump of matter (object) that can be observed, and calculations of the speed of the moving clumps are made without proper account of the fact that when the clump was observed last year it was farther from Earth than during a similar observation this year, then an error is made giving an apparent speed that is greater than the actual speed. The appearance of this type of motion is *across our line of sight*, that is, in the plane of the sky. We must correct our time measurements for the difference in signal transit times to obtain the actual speeds.

To illustrate how this apparent, seemingly impossible speed arises, consider Figure 6-9. Here the light-emitting object was first observed at point A. Unknown to us, it is moving at 0.894 c toward point B (moving mostly toward Earth). In 2.5 years the light emitted at Point A will have traveled 2.5 light years toward Earth as shown by the arc which is part of a light sphere centered on

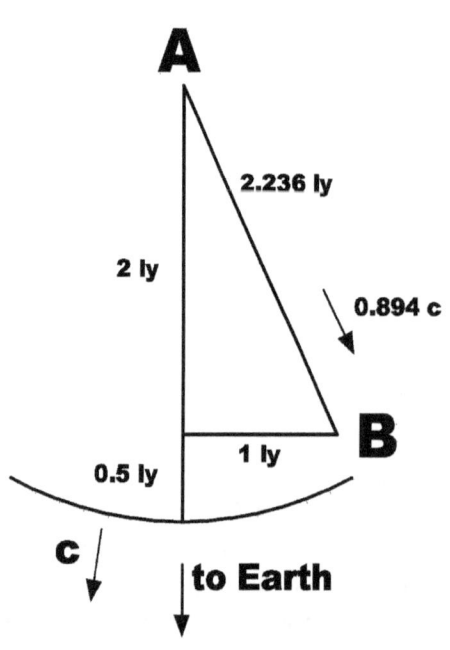

Figure 6-9. An example of superluminal speeds seen across our line of sight.

point A. During this 2.5 years the light-emitting object will have moved 2.236 light years (2.5 years ×0.894 c) to point B, which is 1 light year across our line of sight and 2 light years nearer to Earth than point A. The light from point A will reach Earth 0.5 years before the light emitted at point B because the light from point A is 0.5 light years ahead of the light emitted at point B in their journeys toward Earth. To observers on Earth it will appear that the object has moved 1 light year across our line of sight in 0.5 years, giving an apparent speed of 2c. When such superluminal speeds were first encountered, some astronomers were briefly puzzled.

6-8 Time Travel

Science fiction that deals with time travel often uses the device of parallel universes in time. Typically in these stories, someone travels to the future or the past and encounters her older or younger self. These stories imply that there must be a near infinite number of parallel universes that are co-existing and are closely linked together. There is no indication that these parallel universes actually exist. If they did exist then we might notice events happening without any identifiable cause. Someone in the past does something that changes something in our time. (Of course, there might be a parallel universe that is mostly filled with unmatched socks.) Quantum mechanics (Remember the 12 decimal place predictions in Section 5-7) indicates that the future is *not* determined, but rather is probabilistic. This means that if quantum mechanics were *not* part of natural law, and *if* we could know the precise configuration of the universe at some moment, then in principle we could precisely calculate what the universe would be doing in the future. But because probability enters into *microscopic* events, and these make up *macroscopic* (large) events, then detailed calculations about the future become increasingly fuzzy as we calculate for farther into the future. This fact is usually expressed quantitatively as the "Heisenberg Uncertainty Principle." The effects of this principle plainly show up in sub-microscopic processes, but then large processes are made up of many microscopic processes, so large processes gradually become uncertain also. A joke sign reads, "Heisenberg may have slept here." Not only can we not calculate exactly what the future is to be like, it appears that the future is not absolutely determined. However, the past *is* fixed. Without these probabilities, free thought would not be possible. So instead of saying to the judge, "The Devil *made* me do it," it would be more correct to say, "Heisenberg *let* me do it."

Microscopic events occur all around and within us. Let us consider the process of an oxygen molecule hitting a piece of iron. The oxygen molecule might chemically combine with the iron, creating iron oxide (rust) or it might simply bounce off. It seems that what actually happens in such a process is not predetermined but is governed by probabilities. For example, quantum mechanics might be able to calculate that there is (say) a 1% probability that oxidation will occur during such a collision, but it is completely open as to what actually happens in any particular collision. In the case here, it is most likely that the oxygen will simply bounce off. You may have noticed that rust is an irregular powdery or flaky substance. These grains or flakes form because of these probabilities of oxidation when the iron was first exposed to the oxygen. The rusting does not form uniformly over the surface of the iron. Similarly, no two snowflakes are the same for probability reasons when the flakes are growing.

Time travel into the future is certainly possible. We are doing it as we sit, but only at the rate of 1 second per second which is essentially the same as most things immediately around us. If we move very fast relative to Earth we can effectively speed this up (as Roamer, the traveling twin did), but at a very great cost in energy (see Section 4-6). Traveling into the future by this mechanism would bring about no problems with causality such as you killing your mother before you were conceived. It would be like going on a long trip and then showing back up (because that is what you did). If parallel universes really do exist, then there could also be causality problems when traveling into the future.

Most of the causality problems, such as those depicted in numerous science fiction stories, are caused by traveling backward in time. For example, hopping into your time machine then traveling backward (the cause) and changing something in the past (the effect) means that the effect preceded the cause. This is a clear violation of causality. It is fortunate that there seems to be no possible way to travel into the past. Things could get really messed up. Stories about this process can certainly be fun, but it *does* seem to be impossible. Also, no signal (information) could be sent backward in time either.

This means that you cannot somehow communicate information about some important invention, sporting event or stock market occurrence to yourself in the past and suddenly become rich in the present.

To avoid the annoying signal-travel times over astronomical distances, science fiction writers often invent something like "hyperspace" where signals can travel much faster than light. If this is ever actually discovered, then we will need to replace c with this "hyper c" in the discussion of causality. Some science fiction dodges this issue by claiming that nothing "natural" can travel faster than c, but some "artificial" signals can. But then, aren't humans part of nature?

Continue to enjoy science fiction but remember that it *is* fiction, and fiction almost always depicts relativity grossly incorrectly. Real relativity is not nearly as weird as in science fiction.

6-9 The Twin Paradox Solution: What Each Twin Sees

Besides traveling in space (but ending up where he started in space), the traveling twin, Roamer, as described in Section 5-4, also travels (in a sense) into the future. We say "in a sense," because each twin travels into the future, but space-faring Roamer travels farther into the future than Earth-bound Homer. Traveling into the future is the reason that the twins are being discussed again at this time-travel point in our study. Figure 5-6 is repeated here so that we can refer to it in the following discussions.

Previously, we took the usual god's-eye view of the situation, where we were very high above the experiment so that every part of the experiment was equidistant from us (the observer). As usual, we did this so we did not need to make any corrections for the times it took for light-speed signals to reach us; we *determined* the situation rather than *saw* it.

In Section 5-4 we described the trip and explained that to Homer (in Earth's frame) the time travel was because Roamer was moving at high speed, thus Roamer's time ran slow to Homer, so during the trip less time passed to Roamer than it did to Homer. The cause of the time travel was different as determined by Roamer: To him, it was Earth that moved away at high speed and back again, so from this alone one might think that Homer should have been the younger upon the reunification of the twins. However, the actual time travel occurred because of the turn-around at the far end of the trip, B to C. To accelerating Roamer, Homer and Earth were effectively very high in a gravitational field, so Homer's time ran very fast during this mid-region of the trip. From B to C

Homer's time was fast enough to Roamer to doubly-cancel (cancel and reverse) the effect of Homer moving very fast in Roamer's rest frame.

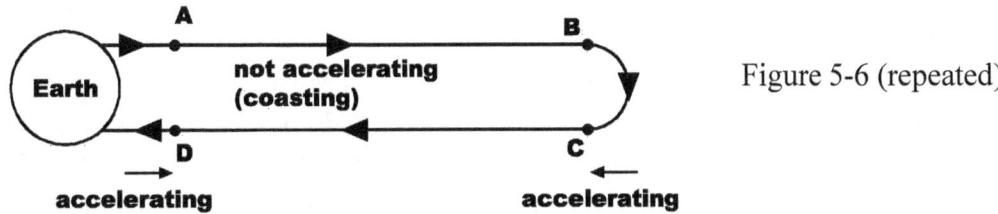

Figure 5-6 (repeated)

Our next description will consider what each twin experiences (feels, sees, etc.). Each twin will send an email to his brother at Christmas; the dates are indicated by a highly accurate clock/calendar on Earth and an identical device traveling with Roamer on the ship. We will assume that this experiment takes place far in the future so that more advanced space vehicles are available and email spam filters have been perfected.

During the two coasting phases of the trip, A to B and C to D, how things appear to each twin will be affected by time dilation because of their relative velocities, and the changing distance between the twins will change the time delay of the light-speed signals between them. Fortunately, the relativistic Doppler formulas take both of these effects into account perfectly. Instead of referring to the frequency of a light wave or photon, we will be referring to the frequency of receiving the "yearly" emails. Realize that the two clock/calendars will usually not be running at the same rate as determined by either twin or us. Also, because the distance between source and observer is changing, the Doppler formulas give an *as seen,* rather than a *determined* result.

We repeat the formula for the relativistic Doppler effect on frequencies (Equation 5-1, derived in Appendix A-14) for reference in the following discussions:

$$frequency_{received} = \frac{frequency_{at\,rest}\sqrt{1-v^2/c^2}}{1+v/c} = frequency_{at\,rest}\sqrt{\frac{c-v}{c+v}} \quad (5-1, repeated)$$

We will use Earth's frame for the following description: Roamer and his dog Rover will travel at a speed of $3c/5$ ($0.6\,c$) on each coasting leg of the trip. Roamer would like to have traveled faster, but taking his large dog along meant that the extra fuel needed for the higher speed had to be left behind. Referring to Table 5-1 in Section 5-2, we see that for this speed on the outward bound leg ($v = +3c/5$), the received frequencies and rate of receiving emails, will be only ½ their rate as sent from the other twin. However, on the return leg ($v = -3c/5$) these received rates will be twice their rate as sent by the other twin. Envision that from A to B the emails sent from Earth are moving at *c*, chasing the ship and gradually overtaking it. On the return journey, the ship will be having head-on collisions with the radio signals (emails) from Earth. These produce the every-other-year, then twice-a-year email arrival rates on the outward and then inward coasting legs of the trip, respectively. How will the *rates* of email reception compare between the twins during either of the coasting periods? They will be exactly the same (except for the content), because we could use either the rest frame of the ship or that of Earth to make the arguments; the A to B and the C to D segments are symmetrical. Corollary 1 assures us that each twin will *determine* that the other is traveling at the same *speed*, but in the opposite direction. The Doppler formulas use the sign convention that *v* is positive if the source and receiver are separating, and negative if they are approaching. Besides the frequency of email reception, a telephone conversation would have words received at ½ the rate that was sent from A to B, and the voices would be lowered to ½ the normal frequency (one octave lower). From C to D, these effects would be twice, rather than one-half, and have one octave higher voices. Email is primarily used by the twins because telephones would present a very irritating time lag in the conversations: Roamer would say something, then it might take years for the signal to travel to Homer. Homer would then reply, but it might take years to reach Roamer. All this time each would have to be holding the phone so as not to miss the message. Smart recording devices help a lot, hence the use of emails, but there would still be the same delays between the emails.

Since each twin experiences the same Christmas email reception frequency as his brother during each coasting phase of the trip, then we might think that each would receive the same total number of Christmas emails. This is *not* true because with time dilation each twin experiences a different duration of these legs of the trip. Their Christmases are not synchronized. In Section 5-4, we saw that it was the acceleration that Roamer experienced at the turn-around that actually produced the time travel (in Roamer's frame). This is because during the turn-around it is *as if* Earth were very high in a gravitational field, which causes time to pass faster on Earth (in the ship's *accelerating* frame). This produces an apparent fast time on Earth. If Roamer looks at Earth between points B and C with a very powerful telescope, he would see activities on Earth speeding up from the half speed for Earthlings that he had been *seeing* on the outward leg of the trip, A to B. Halfway between B and C the ship has momentarily stopped, relative to Earth. The emails from Homer are now arriving at the sent rate of one per Earth year. Through the telescope, events on Earth that occurred some years after Roamer blasted off are now seen at normal speed. That number of years after blast off of the email delay will simply be the number of light years that Roamer has traveled away from Earth. He is at rest relative to Earth at this point, so the only effect is the time it takes for a light-speed signal from Earth to reach him. The *rates of increase* in email arrivals and the apparent speeds of Earthly events will continue until the ship reaches point C where it has attained a speed of $-3c/5$. The engine will then be shut off, and the coasting leg of the return journey, C to D, begins.

A physical explanation of these effects is perhaps the most convincing: The emails are being sent as photons. Even radio signals are photons. We saw in Section 5-3 that as photons move upward in gravity, they do not slow down, but their energy (hence mass) decreases, thus their frequency decreases. Conversely, if photons fall under gravity, they do not speed up, but their energy increases (as does their masses), and their frequency increases proportionally. For the email messages sent from Earth and received by Roamer, their frequency of being sent was controlled by a clock on

Earth. By Postulate 3, acceleration and gravity produce the same effects. Thus, from B to C, as Roamer accelerates toward Earth, clocks on Earth and email arrivals from Homer will be seen by Roamer to gradually increase their rates. This is because Roamer has stopped running away from the stream of emails from Earth and is gradually starting the head-on reception of the signals.

On Earth, Homer and Houser will essentially see the ship move away from Earth, reverse course, and then come back again. It has already been mentioned that they will receive emails much as Roamer and Rover did, but they will not receive as many emails as Roamer and Rover did during the experiment because Roamer never wrote and sent as many. From Section 5-4, less total time, thus fewer Christmases have passed to Roamer than to Homer. Two additional apparent motions will be seen by each twin. From A to B, Table 6-1 in Section 6-7 shows that the $3c/5$ speed of recession will be *seen* by each twin as receding at only $3c/8$. From C to D, the $-3c/5$ speed of approach will be *seen* by each twin as the superluminal speed of $-3c/2$.

We have looked at two paradoxes in this book: the Stick and Slot, and the Twin Paradox. Besides the usual relativistic effects, causality was required to resolve the Stick and Slot paradox, and gravity was needed to resolve the Twin Paradox. There are other relativity paradoxes, but as with these two, the paradoxes resolve themselves when we take all of the relativity effects into proper account.

Epilogue

Now that you know quite a lot about relativity, and hopefully it does not now seem so mysterious, you can appreciate the dictionary definition below:

rel-a-tiv-i-ty (rel′ ə tiv′ i te)   n. a theory formulated essentially by Albert Einstein, that all motion must be defined relative to a frame of reference and that space and time are relative, rather than absolute concepts: it consists of two principal parts. The theory dealing with uniform motion (special relativity) is based on the two postulates that physical laws have the same mathematical form when expressed in any inertial system, and the velocity of light is independent of the motion of its source and will have the same value when measured by observers moving with constant velocity with respect to each other. Derivable from these postulates are the conclusions that there can be no motion at a speed greater than that of light in a vacuum, mass increases as velocity increases, mass and energy are equivalent, and time is dependent on the relative motion of an observer measuring the time. The theory dealing with gravity (general relativity) is based on the postulate that the local effects of a gravitational field and of acceleration of an inertial system are identical. (*Webster's New Universal Unabridged Dictionary*, 1996, Barnes & Noble Books)

(As good as this definition is, it has a flaw. Try to find it. Hint: See Section 1-5. The answer is in Appendix 18)

If you have reached this point, and as would be quite normal, do not completely understand the material at certain points, then armed with the additional experience acquired from the later material, going back and rereading might now make the difference needed for complete understanding. This discussion of relativity has been kept as brief as possible, while still covering the basics. The physics content has also been kept to the minimum required. If you do go back and reread, taking the time to reflect and make sure the material is understood, then you should have a really good start on any other treatment of relativity. In any case, rereading is almost always a good idea for any material that is technical in nature.

If you did not look at the algebraic manipulations in the appendices, this might be a good opportunity to brush up on these skills or to learn some of them, whichever the case may be.

An interesting exercise for the confident reader is to try a different hypothesis and go through the early steps of the development of special relativity. For example, we could keep the first postulate as is, but use the postulate (the new Postulate 2) that light moves at c relative to its source (this includes moving at c relative to objects that it reflects off). If this is done, then Theorem 1 becomes the same as the new Postulate 2. Corollary 1, which depends only on Postulate 1, remains unchanged. When we try the light clocks, the parallel clock (Section 1-15) is the simpler to analyze (assume no length contraction). The light moving to the right has a speed of $c + v$, and a speed of $c - v$ to the left. This will give the result that the clock runs at the same rate regardless of its speed, v , that is, no relativistic effect. The transverse clock (Section 1-9) is now the more complicated. The speed of the light that bounces between the mirrors will have a vertical component of c but will have a horizontal component of v (depending on exactly what the new Postulate 2 says). This will give a total speed of the light that bounces between the mirrors of $\sqrt{c^2 + v^2}$. This is greater than c. We do get self consistency to this point in that this clock also runs at the same rate as the parallel light clock, that is, it does not depend on the velocity, v. Section 1-13 will produce no length contraction, and Section 1-12 will give the same result as before, that is, transverse lengths are absolute. To synchronize clocks by the "flash in the middle" method we would have to specify that the velocity of the flash apparatus was zero relative to the two clocks. The clocks will then be synchronized in all frames. This is as far as the author has taken this pseudo-relativity. Causality will certainly become more complicated, and the Light Spheres Theorem will not be as powerful. I hope that the reader does not ask, "If the results are simpler, and are self consistent, why don't we take this as the truth about the universe?" The answer is, of course, that *these* results *do not* agree with experiments – the arbiter of all theories on the workings of the universe.

However you use this book, it is the author's hope that you will gain enough understanding of relativity so that some criticality can be applied when you hear something about it or when you encounter a science fiction story that supposedly deals with some aspect of relativity. It is completely permissible to suspend disbelief while encountering science fiction. Just be glad that the universe is less crazy than depicted in some science fiction.

If you have read this far, then "Congratulations." I still recommend a second reading; it will go much quicker than the first.

Appendices

Mathematics Lessons:

Note: Simple math lessons are included in the appendices that will allow the reader to be comfortable with the math used in this book. Much of the space in these appendices is devoted to showing the algebraic manipulations that are necessary to simplify expressions or equations that are developed via the gedanken experiments. This will provide many examples of the math learned in the lessons. The steps in the manipulations are shown along with a description of what mathematical steps were taken. Where some *expression* is being manipulated there may be a train of equal signs. Where an *equation* (or inequality) is being manipulated the steps might be separated by the $\Rightarrow$ sign. The *descriptions* use *semicolons* to separate the steps to go from one equal sign or one arrow to the next, while *commas* are used to separate the individual small steps. Examples: (for an equation) . . . ; multiply both sides of the equation by c, then add v/c to each side; . . .   (for an expression) . . . ; multiply numerator and denominator by c, then take the square root; . . . Depending on the situation, the verbal explanation might be written above or below the math steps.

Contents:

Details of Mathematical Manipulations, Supplementary Topics, and Answers to Questions
   Contents:

Appendix 1: The Fundamental Principle of Algebraic Manipulations, Working with Fractions, Units, Scientific Notation and Prefixes

A1-1 The Fundamental of Algebraic Manipulation

If an equation is true, then it will remain true if we: Add the same thing to each side, subtract the same thing from each side, multiply or divide both sides by the same thing, square or square root both sides, or do any other operation to both sides. For example: Suppose that we have the equation $1+1=2$. Let us multiply each side of this equation by 3. This will produce the equation $3+3=6$, an also correct equation. Notice that we must multiply each term by the multiplier (a "term" is an added or subtracted item in an equation). More on this later. If we now add 4 to each side of last equation we obtain $3+3+4=6+4$. Both sides of this equation equal 10, so the equation is still true.

A1-2 Working with Fractions

Suppose that an orange has twelve equal segments. We could refer to one segment as "one twelfth of the orange." We write this as "1/12 orange." The slash (/) is a fraction bar which signifies division. One segment is 1 orange divided by 12. If you gave 5 segments to a friend, then he would have five twelfths (5/12) orange. The first or upper number is called the "numerator" of the fraction, and the second or lower number is called the "denominator."

Fractions may be greater than one. For example, suppose that we had 18 orange segments. this would be eighteen twelfths of an orange (18/12 orange). We will see below that this is 3/2 orange, or one and one half orange.

A1-2.1 Adding or Subtracting Fractions:

Suppose that you must add 1/4 to 5/4. You simply get $\dfrac{1}{4}+\dfrac{5}{4}=\dfrac{1+5}{4}=\dfrac{6}{4}$

This is easy because both fractions have the same denominator, namely 4. We are simply adding one "fourth" to five "fourths." To add fractions with different denominators, one or both of the fractions must be converted to an equivalent fraction(s) where each has the same denominator. Here is an easy way to convert fractions so that they have equal denominators: The value of a fraction remains unchanged if we multiply (or divide) a fraction's numerator and denominator by the same number. For example, let's multiply both numerator and denominator of the fraction 1/3 by 2. The fraction then becomes 2/6, which is clearly the same value as 1/3. Now let's add the fractions 1/2 and 2/3, but what should we multiply the numerator and denominator of each fraction by? The easiest way is to simply multiply the numerator and denominator of each fraction by the

126

denominator of the *other* fraction. So we would convert 1/2 to 3/6, and convert 2/3 to 4/6. These then easily add to

$$\frac{1}{2}+\frac{2}{3}=\frac{3}{6}+\frac{4}{6}=\frac{3+4}{6}=\frac{7}{6}$$

If we must add a whole number to a fraction, the whole number must first be written as a fraction. For example, if we must add 3 to 1/2, then convert 3 to 3/1 and proceed as above to obtain 7/2.

To subtract fractions all that is different is to replace the + sign with the − sign and do the indicated subtraction rather than addition. It often happens, as in our first example, that the answer fraction can be simplified. In this case 6/4 can be converted to 3/2 simply by dividing both numerator and denominator by 2.

In the manipulations in this book, we will usually be using letters to represent numbers. There are several reasons to do this, but the most important reason is so that we can end up with a formula where we can then insert whatever numbers that are needed. Without doing it this way, we would have to repeat the manipulations each time we needed the calculation done with different input numbers.

Suppose we need to subtract c/d from a/b, where these four letters represent numbers. We simply proceed as above except that we represent multiplication by writing the letters side-by-side. We would then write $\dfrac{a}{b}-\dfrac{c}{d}=\dfrac{ad}{bd}-\dfrac{cb}{bd}=\dfrac{ad-cb}{bd}$. (We multiplied both numerator and denominator of the first fraction by *d*, and numerator and denominator of the second fraction by *b*.) When adding or multiplying numbers, they can be written in either order. That is, $a + b = b + a$ and $ab = ba$. This is not true for subtraction or division.

We will make considerable use of the equation "distance equals rate times time," that is: $D = v\,t$. (*v* stands for "velocity" or "speed" or "rate.") We will also need two other forms of this equation. Divide both sides by *t*, turn the equation around, and we obtain $v = D / t$ which says, "Velocity equals distance over (divided by) time." This is like miles per hour. If instead of dividing by *t* as we did above, we instead divide both sides by *v* and turn the equation around, we would obtain $t = D / v$ which says, "Time equals distance over speed (or velocity or rate). Example: How long would it take to drive a distance of 120 miles at the rate of 60 miler per hour? $t = 120$ miles / 60 miles per hour = 2 hours. ("per" means "divided by" or "/ ")

A1-3 Units:
The way the units work in fractions is the same as numbers. Continuing the last example:

$$2\,\frac{miles}{miles/hour}=2\,\frac{1}{1/hour}=2\ hour(s)$$

where we divided numerator and denominator by "miles"; then multiplied the primary numerator and primary denominator by *"hour"*. *"Hour"* and *"hours"* have the same dimensions, as do the singular and plural forms of all other units (or dimensions).

A1-4 Scientific Notation (also called "Exponential Notation"):

When a number has a very large or a very small magnitude, such as 1230000000000., it is convenient to write it in a notation such as: 1.23×10^{12}. 10^{12} means twelve 10's multiplied together, or a one followed by twelve zeros. 1.23×10^{12} also means to start with 1.23 and shift the decimal point to the *right* twelve spaces, filling in the needed zeros. This notation prevents us from having to count the number of zeros in the number; there are also other advantages: A very small number such as 0.00000003 can be written as 3×10^{-8}. The negative exponent means to shift the decimal point to the *left* eight spaces, filling in the needed zeros. 10^{-8} also means $1/10^{8}$, that is, one divided by 10^{8}. A negative 567.0 can be written as -5.67×10^{2}.

If we must multiply two exponential numbers, we multiply the decimal portions (these are called "mantissas") and add the exponents together. For example: 2×10^{6} times $3 \times 10^{-2} = 6 \times 10^{4}$. To divide, we divide the decimal portions and *subtract* the exponents. For example:
$3 \times 10^{-2} / 2 \times 10^{6} = 1.5 \times 10^{-8}$. (Negative 2 minus 6 equals negative 8.)

A1-5 A List of Commonly Used Prefixes on Units:

Name	abbr.	Multiplier	In words
nano-	n	$\times 10^{-9}$	one billionth
micro-	μ	$\times 10^{-6}$	one millionth
milli-	m	$\times 10^{-3}$	one thousandth
centi-	c	$\times 10^{-2}$	one hundredth
kilo-	k	$\times 10^{3}$	1000 times
mega-	M	$\times 10^{6}$	one million times
giga-	G	$\times 10^{9}$	one billion times

Appendix 2: Binomials, Factoring, and Solving Simple Equations
A2-1 Binomials

A binomial is two numbers that are either added or subtracted. Usually the term is used only when we are using letters to represent numbers so that we cannot just go ahead and add or subtract them. So $a + b$, $a - b$, $-a + b$ or $-a - b$ are binomials. A common binomial used in relativity is $1 \; v^2/c^2$. The superscript 2 on v means that v is "squared," that is, multiplied times itself (like in the area of a square). It is similar for c^2. A trinomial is similar to a binomial but has three numbers, or letters, to be added or subtracted. The items to be added or subtracted are referred to as *terms*. *Polynomials* are like binomials but with more than two terms to be added or subtracted.

A2-1.1 Adding or subtracting binomials

If your snack bag contains one apple and two oranges, then we could write the binomial 1 apple + 2 oranges. If your friend's snack bag contains an orange and a candy bar, then we could get the total quantity of snack supplies by simply adding: (1 apple + 2 oranges) + (1 orange + 1 candy bar), but we would usually combine like terms, so the sum would be 1 apple + 3 oranges + 1 candy bar. Another example of adding binomials could be $(a + 2b) + (3a - b) = (4a + b)$. An example of subtracting binomials would be $(a + 2b) - (3a - b) = (a + 2b) + (-3a + b) = -2a + 3b$. That is, when subtracting binomials, simply change both signs in the binomial to be subtracted, then add.

A2-1.2 Multiplying binomials

Suppose that we had $(a + 2b)(-3a + b)$. Simply multiply the first term in the first binomial times each term in the second binomial, then multiply the second term in the first binomial times each term in the second binomial, then add up all four products. That is $(a + 2b)(-3a + b) = -3a^2 + ab - 6ab + 2b^2 = -3a^2 - 5ab + 2b^2$. Remember the rule of multiplying (or dividing) signed numbers, "Two like signs multiplied or divided gives a positive, two unlike signs gives a negative." A common product is $(c - v)(c + v) = c^2 - v^2$. Verify this for yourself.

Usually there is little that can be done when dividing binomials unless the two binomials are equal or one is a multiple of the other. (But see "Factoring" below, Section A2-2.)

A2-1.3 Squaring binomials

It often happens that a binomial must be multiplied times itself. For example, $(3a + b)^2 = 9a^2 + 6ab + b^2$, a trinomial. One may either write the binomial beside itself and multiply as above, or one may remember the rule, "Square the first, plus two times the first times the second, plus the second squared." Here are some more examples:

$(3a - b)^2 = 9a^2 - 6ab + b^2$ (remember + times − is a −)
$(a - b)^2 = a^2 - 2ab + b^2$ (It is *not* $a^2 - b^2$)
$(a + b)^2 = a^2 + 2ab + b^2$ (It is *not* $a^2 + b^2$)

Sometimes one needs to recognize these forms in reverse. For example, we may need to convert $4a^2 + 4ab + b^2 = (2a + b)^2$.   (This curious form comes about because 2 plus 2 = 2 times 2.)

A2-2 Factoring

Factoring is the reverse, not opposite, process of the multiplying that we did above. For example: If we have 4 + 6 we could write this as 2 (2 + 3). We would say, "Factoring out a 2." When we factor out a number from a binomial or polynomial, we must divide each term in the binomial by that factor to keep the expression equal to what we had before. A common factoring operation that occurs in relativity is the following: $c^2 - v^2 = c^2 (1 - v^2/c^2)$ where we have divided each term by c^2 , and written c^2 as multiplied times the *quotient* (the result of a division). Visualizing the c^2 multiplied back times the last binomial shows that the equality still holds.

A2-3 Solving Simple Equations

Equation 6-1, $LR + vt = ct$, is an example of common equations in the study of relativity. We need to find what t is so that it will satisfy this equation (that is, "solve for t"). We will show two methods, each of which will use the principles from Appendix 1, and the earlier sections of Appendix 2.

Method 1: (below) Start with Equation 6-1; subtract vt from each side (This gets all the t's, the desired unknown, on the same side of the equation); factor t from each term on the right side; divide each side of the equation by $c - v$; re-write the equation in reverse order (turn it around).

$$LR + vt = ct \implies vt - ct = -LR \implies ct - vt = LR \implies t(c - v) = LR \implies t = \frac{LR}{c - v}$$

Method 2: (below) Start with Equation 6-1; subtract both LR and ct from each side; multiply both sides by -1 (This removes the minus sign on the right side.), rearrange the order on the left side; factor t from each term in the left side; divide both sides by $c - v$.

$$LR + vt = ct \implies LR = ct - vt \implies LR = t(c - v) \implies \frac{LR}{c - v} = t \implies t = \frac{LR}{c - v}$$

As long as the principles of algebraic manipulation (if one: adds the same thing to each side, subtracts the same thing from each side, multiplies or divides each side by the same thing), the equation remains true. The equation also remains true if any single operation is performed on each side of the equation. These include such operations as: squaring both sides, square-rooting each side, and any similar thing. Also, the numerator and denominator of any fraction may be multiplied or divided by any quantity without changing the value of the fraction. Appendix 7 contains some more examples of the common procedure shown above. As a starting goal in solving an equation, it is essentially always better to work toward getting all instances of the unknown quantity, t in the above case, on the same side of the equation.

Appendix 3: Square Roots, Imaginary and Complex Numbers, and Mathematical Logic
A3-1 Square Roots

"Squaring" means multiplying a number times itself. For example, 3 squared (written 3^2) = 9. Notice that negative three (-3) squared is also 9 because the product (or quotient) of two negative numbers is also a positive number.

Taking a "square root" is the opposite of squaring. The square root of 9 is 3. Actually, it could also be -3 as well. More about this later. The square root operation is usually designated by a symbol called a "radical." It looks a little like the symbol for long division: $\sqrt{9} = 3$. In all the applications in this book we will not need the negative square root, but there are many cases where

130

one might need it. By definition, the radical stands for the positive square root. If a situation dictates that we desire both the positive and the negative square root, then we place the ± (plus or minus) symbol in front of the square root. In this case we would write $\pm\sqrt{16} = \pm4$.

For some applications in this book, we will have need to square a square root, or find the square root of something squared. In both cases, the two processes simply cancel each other. Examples:

$$\left(\sqrt{a+b}\right)^2 = a+b \qquad or \qquad \sqrt{\left(1-v^2/c^2\right)^2} = 1-v^2/c^2$$

In the second example directly above, if we would have accepted either the positive or the negative square root, we would have placed the ± symbol in front of the radical as well as in front of the entire right side.

Note: The final three sections of Appendix A3 are not needed for this book. However, a slight mention is made in the body of the book about imaginary numbers.

A3-2 Imaginary numbers:

In the examples above we only took square roots of numbers that are positive. What would we do to take the square root of a negative number? At first it seems that we can never take a square root of a negative number because a negative number squared, like a positive number squared will give a positive number.

We define the square root of a negative number to be "imaginary." One might wonder about what use there might be for imaginary numbers, but as it turns out, there are many uses. More on this later.

For convenience, mathematicians have defined the square root of −1 to be designated by the symbol, "i" (for imaginary). That is $\sqrt{-1} \equiv i$. The triple bar symbol means "is defined as." The i symbol makes it easy to write the square root of a negative number because $-25 = (25)(-1)$. It is true that the square root of a product (or quotient) is the product (or quotient)[1] of the square roots. So we may write:

$$\sqrt{-25} = \sqrt{25} \cdot \sqrt{-1} = 5i$$

---

[1] This is only true for a fraction (a quotient) if the denominator is positive. If the denominator is negative, then first multiply numerator and denominator of the fraction by −1 so that the denominator becomes positive, then take the square roots. This situation does not arise often and is a little-appreciated mathematical fact.

There are only two mentions of i in the body of this book, and there it is only needed for an additional justification of a minus sign.

A3-3 Complex numbers

(This book will have no use of these; this is for information and practice only. There is an interesting bit of trivia near the end of sub-section A3-4.)

A complex number is the sum (or difference) of a real number and an imaginary number. For example: $3 + 4i$ .These are handled the same way as binomials of real numbers. The only thing to remember is that when i^2 appears, it is simply replaced by -1 . For example:

$$(3+4i)\cdot(5-2i)=3\cdot5-3\cdot2i+4i\cdot5-4i\cdot2i=15-6i+20i-8i^2=$$
$$15+14i-8\cdot(-1)=15+14i+8=23+14i$$

A3-4 Applying mathematics to nature:

The main reason for bringing up complex numbers is that they do behave in the way that some physical processes do. The primary reason that we use numbers, and mathematical operations, is that mathematics does emulate (behave the same way as) nature. A trivial example is if we place three oranges in a bag, then place two more oranges in the same bag, then we now have five oranges in the same bag. We see that addition (a part of mathematics) does emulate this simple process in nature. More complicated physical situations require more complicated math. It turns out that certain things in nature really do behave in the way that complex numbers behave. One example is from the alternating current (AC) electricity that we use. The magnitudes (volts or amps) and their phase angles [a measure of just when (say) one voltage peaks relative to the peaking of the other voltage], can be expressed using complex numbers. Complex numbers used in this way are called "phasors," a name that has been borrowed for the ray guns used in a famous science fiction TV series.

Another reason to use mathematics is that we can better keep up with certain logical manipulations when we can write them down and then use trusted mathematical manipulations. This occurs many times in this book. It turns out that the reasoning done in relativity is intricate enough that it is not feasible to do the logic without this mathematical assistance, even the way that it is done here using gedanken experiments.

Appendix 4: The Pythagorean Theorem, Trigonometric Functions, and Vector Addition / Subtraction

A4-1 The Pythagorean Theorem in 2-D

The Pythagorean Theorem has been known for over 2000 years. This well-known theorem of geometry states that for a right triangle (One angle is 90°, and is marked by the square in Figure A4-1, except for the middle triangle for clarity.), the hypotenuse (the side opposite the right angle) squared is equal to the sum of the squares of the other two sides. (Here, c is *not usually* the speed of light, but all three sides of any triangle must have the same units.) That is:

$$c^2 = a^2 + b^2$$

A simple right triangle that is used several times in the body is the 3, 4, 5 triangle. It is simple to show that these side lengths do form a right triangle because $3^2 + 4^2 = 5^2$ $(9 + 16 = 25)$.

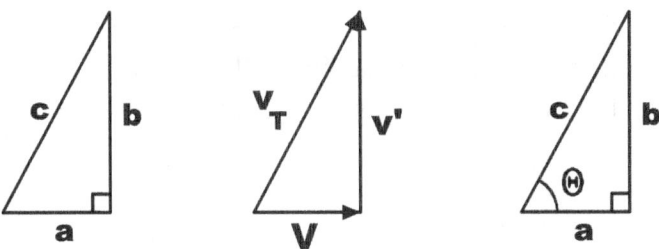

Figure A4-1

The middle triangle shows the case where we are not dealing with a simple triangle of lengths, but a triangle made up of velocities. Here we have:

$$v_T^2 = v'^2 + V^2$$

as it is used in Appendix 16. Still using the middle triangle, we note that the vector, V, with the vector v' placed tail-to-head on it equals the (Total or sum) vector v_T. This is vector addition. We can add vectors by this tail-to-head placement. Since V and v' are at right angles, then we can call this pair the "components" of v_T. This concept is first used in Section 2-4, and is further developed in Section A4-4.

The third right triangle is like the first, but one of the angles (not the right angle) has been labeled with the Greek letter θ (theta). By definition, the "sine of θ" is $b\,/\,c$, the "cosine of θ" is $a\,/\,c$, and the "tangent of θ" is $b\,/\,a$. These functions are usually abbreviated by their first three letters, thus we write (The $\equiv$ symbol means "is defined as."):

$$\sin\theta \equiv \frac{b}{c} \qquad \cos\theta \equiv \frac{a}{c} \qquad \tan\theta \equiv \frac{b}{a}$$

We have little need for these trigonometric functions in this book, except that in Section 6-1, the arcsine function is briefly used. An arcsine function can be read as "the angle whose sine is" (say) 0.5. Thus, the arcsine is the angle whose sine function is 0.5 in this case. In Section 6-1, we used the arcsine of 4/5. This is 53.13°. That is, sin (53.13°) = 4/5. The other arc- functions are similar. The arc- functions are also written as: $\sin^{-1}$, $\cos^{-1}$, and $\tan^{-1}$.

A4-2 The Pythagorean Theorem in 3D

Figure A4-2 shows a three dimensional extension to the normal (2-D) Pythagorean Theorem.In this case we have $x^2 + y^2 + z^2 = R^2$. This extension to 3-D can be seen to be true by drawing a line in the x-z plane from the origin to the point directly under the event. This line will have a length of $r = \sqrt{x^2 + z^2}$ according to the 2-D Pythagorean Theorem. This diagonal line forms another 2-D right triangle that stands on this diagonal line in the x-z plane, and whose othersides are the vertical line of length y and the radius, R, as its hypotenuse. Again using the 2-D Pythagorean Theorem, we obtain:

$$R^2 = r^2 + y^2 = \left(\sqrt{x^2 + z^2}\right)^2 + y^2 = x^2 + z^2 + y^2 = x^2 + y^2 + z^2$$

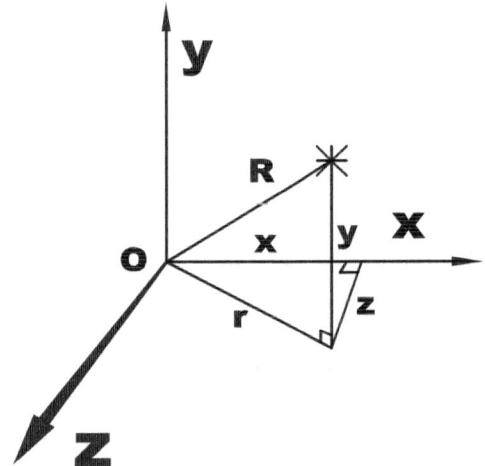

Figure A4-2 The 3-D Pythagorean Theorem. The lines marked x, y, and z are parallel to the X, Y, and Z axes.

which completes the proof. We squared the square-root, then rearranged the terms in the sum.

A *spherical surface* whose center is at the origin will be formed from all points with coordinates (x,y,z) that satisfy the equation:

$$x^2 + y^2 + z^2 = R^2$$

where here R is the radius of the sphere (not the Relativity Factor). This is because a sphere has a constant radius. That is, all points on the sphere are the same distance from the center.

For a light sphere (see Section 3-2) that has been growing for a time t, we can write:.

$$R = ct \Rightarrow R^2 = c^2 t^2 \Rightarrow R^2 - c^2 t^2 = 0$$

A non-zero right side merely means that the light sphere started growing earlier or later than $t = 0$. Appendix 17 uses this.

A4-3 Question from Section 1-15 on the moving angled line:

The hint said to consider the angled line to be the hypotenuse of a right triangle. See Figure A4-3. The side of the triangle that is parallel to the motion will contract by the relativity factor, R, but the side perpendicular to the motion will remain unchanged. Hence, any line will be more perpendicular to the motion when it is viewed from a frame in which the line is moving.

Notice that any line that is perpendicular to the relative motion between the two frames will remain completely unchanged from one frame to the next. A line that is parallel to the motion will not be changed in direction, but its length will be contracted by the relativity factor, R.

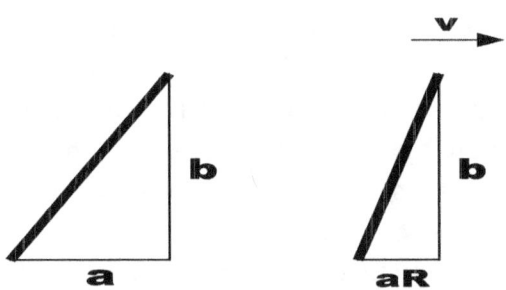

A4-4 Vectors: Adding and Subtracting them, and Vector Components

A "vector" has a direction in space and a "magnitude." Section 4-1 gives some examples of vector quantities. The easiest vectors to visualize are "displacement" and "velocity." For example, if you move from point A to point B by whatever route, then the displacement vector will have the direction from A straight toward B. Its magnitude will be the straight-line distance between A and B and will have the units of kilometers, miles, etc. If

Figure A4-3 An angled line at rest (left), and moving to the right with a speed v (right)

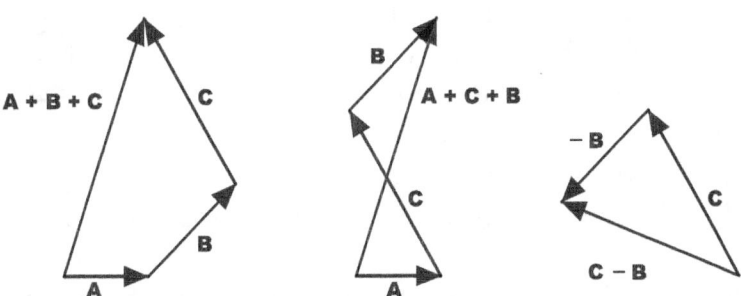

Figure A4-4. The left two figures show two ways that vectors **A**, **B**, and **C** can be added using the "tail to head" method. The middle figure shows that the same result is obtained regardless of the order of adding the vectors. The right figure shows Vector Subtraction. In this case, $-$ **B** is added to **C**, giving **C** $-$ **B**.

something is moving at sixty miles per hour northward, then the velocity magnitude is sixty miles per hour, and the direction of the vector is pointed northward. The *magnitude* of a *velocity* vector is called "speed." Other common examples of vector quantities are forces (the pull of a rope or of gravity, and the push of a boxing glove. The most important vector quantities in relativity are velocities and momentum. Section 4-1 describes momentum. Vectors are usually represented by arrows drawn on the paper. The direction of the arrow represents the direction of the vector, while the length of the arrow represents its magnitude.

Vector addition is easy to visualize: One starts with one the vectors to be added and then places the tail of the next vector to be added at the head of the *immediately previous* vector. This continues until all the vectors that are to be added have been so placed. The sum of these vectors is the arrow whose tail is at the tail of the first vector to be added, and whose tip is at the tip of the final vector that was added. Figure A4-4 shows examples of this process. Notice that, as in ordinary addition, the order that the vectors are added does not affect the final result. To subtract vectors, follow the same procedure as in addition, except reverse the direction(s) of the vector(s) to be subtracted. See Figure A4-4.

Sometimes it is convenient to "resolve" a vector into its "components." As shown in Figure A4-5, this process is simply to define two or three mutually perpendicular directions. Here we have used "horizontal" and "vertical." The two components are simply the projections, or shadows of the vector when a light is imagined to shine perpendicular to that axis (direction). Figure A4-5 shows the vector **B** from Figure A4-4 resolved into its horizontal and vertical components. The light could shine from the right to give us $\mathbf{B}_v$. The light could shine from above to give us $\mathbf{B}_h$. Notice that the vector **B** is the (vector) sum of its components, that is $\mathbf{B} = \mathbf{B}_h + \mathbf{B}_v$. This can be seen if we imagine $\mathbf{B}_v$ being slid to the right until its tail reaches the tip of $\mathbf{B}_h$.

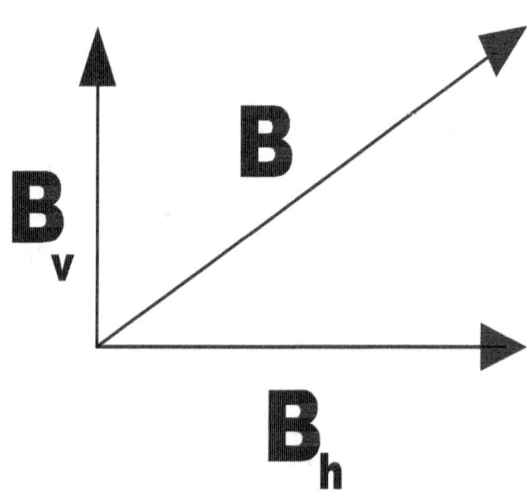

If several vectors have been resolved into their components (must have used the same directions for the axes), then vector summation /

Figure A4-5. The **h**orizontal and **v**ertical components of the vector **B** are shown. These two vectors may be added (vectorially) to give back the original vector **B**.

subtraction can be accomplished by summation / subtraction of the respective components. For example, add all the horizontal components; this will be the horizontal component of the summed vectors. Then do the similar operation for the vertical components.

Appendix 5: The Exponential Function and Comparable Other Functions

We might define the "function", $f(x) = x^2 + 3$. What this means is that if we have a number, say 5, and we want to know what this particular function "f of 5" is, then from the definition of this function we would simply square the 5, giving 25, then add 3, giving $f(5) = 28$. There are many uses for functions. The square root operation is a very common function, and our relativity factor is a function of velocity.

The "exponential function" is written as e^x, where e is the number 2.7182818 ..... (as with the number π, e cannot be completely written as a decimal or a fraction). This function uniquely has the property that its slope (rise over run, or its rate of increase) is equal to its value for all values of x. While we can never completely write down a formula for obtaining the exact value of the exponential function for a certain value of x (except for $x = 0$. Any non-negative number raised to the zero power equals 1.). We can give an expression that allows us to calculate the exponential function to any desired accuracy. It does take increasingly more labor to obtain ever-smaller increases in precision. The expression for the exponential function is in the form of an infinite series.This is summing a list of numbers whose valueseventually get progressively smaller and smaller. This series is:

$$e^x = 1 + x + \frac{x^2}{2!} + \frac{x^3}{3!} + \frac{x^4}{4!} + \ldots$$

where $3! = 3 \times 2 \times 1 = 6$, and $4! = 4 \times 3 \times 2 \times 1 = 24$, etc. $3!$ is called "3 factorial," etc. $x^3 = x \cdot x \cdot x$, that is, x multiplied times itself three times, etc.

The exponential function is always available on scientific calculators and in programming languages for computers. In fact, the series above is what is used to calculate this function when a value is entered and the button is pushed. The larger the value of x, the more terms that must be included in the summation to achieve some particular level of accuracy, say 12 decimal places.

Because its rate of increase is always equal to its value, the exponential function is useful in describing any situation where the rate of growth of something is proportional to the amount of that something. Population is a good example. If all other things are the same, then the number of babies born will be proportional to the number of people, thus this function describes population growth. The term "exponential growth" comes from this function.

In Section 5-6, the exponential functions are mentioned. In this case, the values of $+2\Phi$ and -2Φ are the values of x that are plugged into the exponential function. Using the series above, we find that:

$$e^{2\Phi} = 1 + 2\Phi + 2\Phi^2 + 4/3\,\Phi^3 + 3/4\,\Phi^4 + \ldots$$
$$e^{-2\Phi} = 1 - 2\Phi + 2\Phi^2 - 4/3\,\Phi^3 + 3/4\,\Phi^4 - \ldots$$

The alternating signs in the second form come from raising a negative number, -2Φ, to odd and even powers. A negative number to an odd power, say 1 or 3, will be negative, while a negative number raised to an even power, say 2 or 4, will be positive.

The two functions that appear in the metric of General Relativity are:

$$\left(1+\Phi/2\right)^4 = 1+2\Phi+3/2\ \Phi^2 +1/2\ \Phi^3 +1/16\ \Phi^4 \quad (\textit{exact, not a series})$$

$$\left(\frac{1-\Phi/2}{1+\Phi/2}\right)^2 = 1-2\Phi+2\Phi^2 -3/2\ \Phi^3 +\Phi^4 - \ldots$$

The first equation immediately above is obtained by simply squaring the binomial in ( ), then squaring again. The second equation is obtained by separately squaring the numerator and the denominator, then doing long division. See an algebra text for the process of dividing polynomials. It is really no different from long-dividing numbers, but at first it does seem different. A brief algebra text is listed in Suggested Readings.

Appendix 6: The Transverse Light Clock Manipulations

By using the Pythagorean Theorem we arrived at Equation 1-1 shown below on the left. We first subtract $(v\, t\, /\, 2)^2$ from each side of the equation. Students will usually say, "Take it
[i.e., $(v\, t\, /\, 2)^2$] to the other side and change the sign."

The next four steps are: Multiply both sides by 4; write the left side as $t^2 (c^2 - v^2)$, this is called *factoring*, see Appendix A2-2 ; divide both sides by $(c^2 - v^2)$; write $(c^2 - v^2)$ as $c^2 (1 - v^2 / c^2)$. This is more factoring, see Appendix A2-2. We factored c^2 out of the binomial.

$$\left(ct/2\right)^2 = \left(vt/2\right)^2 + L^2 \ \Rightarrow\ \left(ct/2\right)^2 -\left(vt/2\right)^2 = L^2$$

(below): Take the square root of each side; remember that $t' = 2L\, /\, c$ (time = distance / rate); Multiply both sides by the square root and rearrange the equation, right-to-left.

$$\Rightarrow c^2t^2 -v^2t^2 = 4L^2 \Rightarrow t^2 \left(c^2 -v^2\right)=4L^2 \Rightarrow t^2 =\frac{4L^2}{c^2 -v^2} \Rightarrow t^2 =\frac{4L^2}{c^2\left(1 -v^2 / c^2\right)}$$

(Very Important!) This last equation is Equation 1-2. This equation says that when a time, t, passes between two events as determined in our (un-primed) frame, then we determine that less time, t', has

$$\Rightarrow\ t =\frac{\dfrac{2L}{c}}{\sqrt{1 -v^2 /c^2}} \ \Rightarrow\ t =\frac{t'}{\sqrt{1 -v^2 /c^2}} \ \Rightarrow\ t'=t\sqrt{1 -v^2 /c^2}$$

passed between these same two events in a frame that moves at a speed, v, relative to us. This square root, the relativity factor, R, is always less than the value one, except it is equal to the value one when $v = 0$. In this case the two frames are NOT moving relative to each other. Zero relative speed simply means that the two frames are one and the same frame.

Appendix 7: The Parallel Light Clock and Clock Synchronization Manipulations
A7-1 Parallel Light Clock

For the Parallel Light Clock, we had (inserting the expression for R in Equation (1-3):

$$t = t_R + t_L = \frac{L\sqrt{1 - v^2/c^2}}{c - v} + \frac{L\sqrt{1 - v^2/c^2}}{c + v} = L\sqrt{1 - v^2/c^2}\left(\frac{1}{c - v} + \frac{1}{c + v}\right)$$

where we factored the identical numerators out of the binomial.

We now will work with the expression within the ( ): (below) We add the two fractions in ( ) by the method in Appendix A1-2 where each fraction's numerator and denominator are multiplied by the denominator of the other fraction – then both fractions have the same denominator; add the numerators; (second line below): add out the v's in the numerator, and multiply the two binomials in the denominator (see Appendix A2-1.2); then factor c^2 from the denominator (see Appendix A2-2); and finally divide both numerator and denominator by c.:

$$\left(\frac{1}{c - v} + \frac{1}{c + v}\right) = \frac{(c + v)}{(c - v)(c + v)} + \frac{(c - v)}{(c - v)(c + v)} = \frac{(c + v) + (c - v)}{(c - v)(c + v)}$$

$$= \frac{2c}{c^2 - v^2} = \frac{2c}{c^2\left(1 - v^2/c^2\right)} = \frac{2}{c\left(1 - v^2/c^2\right)}$$

(below) Putting the whole expression back together; grouping the 2 and the L, and dividing the numerator (out front) and denominator by the square root (The parentheses is the square of the relativity factor); finally recognizing $2L/c$ as t':

$$t = L\sqrt{1 - v^2/c^2}\left[\frac{2}{c\left(1 - v^2/c^2\right)}\right] = \frac{2L}{c\sqrt{1 - v^2/c^2}} = \frac{t'}{\sqrt{1 - v^2/c^2}}$$

Now we multiply both sides by the square root, R, and turn the equation around, right for left, we have:

$$t' = t\sqrt{1 - v^2/c^2} \qquad\qquad (1\text{-}2)$$

A7-2 Clock Synchronization

We had the first equation below; then, factor out t_2 from the left side (see Appendix A2-2); then divide both sides of the equation by $c + v$.

$$ct_2 + vt_2 = \frac{DR}{2} \implies t_2(c + v) = \frac{DR}{2} \implies t_2 = \frac{DR}{2}\,\frac{1}{c + v}$$

In the equation below: Subtract vt_3 from each side; factor out t_3 from the right side, and turn the equation around; divide both sides by $(c - v)$.

$$vt_3 + \frac{DR}{2} = ct_3 \implies \frac{DR}{2} = ct_3 - vt_3 \implies t_3(c - v) = \frac{DR}{2} \implies t_3 = \frac{DR}{2}\,\frac{1}{c - v}$$

We now subtract t_2 from t_3 and factor $DR/2$ from each term.

$$t_3 - t_2 = \frac{DR}{2}\left(\frac{1}{c - v} - \frac{1}{c + v}\right)$$

For the final step, the manipulation is similar to that for the Parallel Light Clock above, with the exception of a 2 in the denominator, a minus sign instead of a plus sign, and there the Relativity Factor was written out from the start:

$$t_3 - t_2 = \frac{DR}{2}\left(\frac{1}{c - v} - \frac{1}{c + v}\right) = \frac{DR}{2}\left(\frac{(c + v) - (c - v)}{(c - v)(c + v)}\right) = \frac{DR}{2}\left(\frac{c + v - c + v}{c^2 - v^2}\right)$$

$$= \frac{DR}{2}\left(\frac{2v}{c^2\left(1 - v^2/c^2\right)}\right) = \frac{DRv}{c^2 R^2} = \frac{Dv}{c^2 R}$$

(First step above): To subtract the two fractions, we used the method shown in Appendix A1-2 where we multiplied the numerator and denominator of each fraction by the denominator of the other fraction. Each fraction would then have the same denominator so we could simply combine them into one fraction; (second step in the first line above): apply the minus sign to each of the terms in the second ( ), multiply the two binomials in the denominator. (second line above): In the numerator the c's add out and the v's combine, factor c^2 from each term in the denominator - this gives us the relativity factor, R, squared; divide out the 2's; divide numerator and denominator by R.

Appendix 8: Solutions to Exercises 2-1 and 2-2, and the WWV method of Clock Synchronization

A8-1 Exercise 2-1 Examine Figure A8-1. The observer is moving upward at a speed, v, perpendicular to the separation between the clocks. In this observer's frame the clocks are moving downward at the same speed, v. In the upper half the clocks are paused at zero. A flash goes off midway between the clocks. In the lower half of the figure the flash reaches both clocks simultaneously (in this frame). The clocks start and are clearly synchronized in this frame. We have not had to state the value of the speed, v, so the clocks will be synchronized in any frame moving at any speed perpendicular to the separation of the clocks. (Light outruns everything.)

Exercise 2-2

The hint said to place a third clock at the appropriate position. It has been placed at the lower left, forming a right triangle, though the upper right corner would have been as good a choice. In Exercise 2-1 we showed that two synchronized clocks that are separated at right angles to the motion between the frames will be determined to be synchronized in both frames. We already know from Equation 2-1 that the lower left clock in Figure A8-2 will be ahead of the rightmost clock by Dv/c^2, so this is what the upper clock will also read. So, D is still the separation between the clocks parallel to the line of relative motion between the two reference frames.

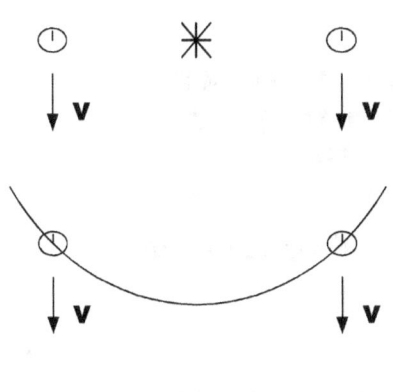

Figure A8-1

8-3 The WWV method of Clock Synchronization:

The WWV method of synchronizing several clocks placed in various positions was mentioned in several sections of the body portion. This description will refer to the three clocks in Section 2-4 whose placement formed a right triangle within a spaceship. One might wonder how the synchronization of these three clocks would be carried out within the ship (in the ship's rest frame). The lower two clocks can be synchronized by the "flash in the middle" method previously described in Section 2-2. To synchronize the upper clock with the one beneath it, the ship's crew can use the "WWV" method. This is named after the radio station that broadcasts the standard time signals to all of North America. The WWV method works like this: The distance, d, between the left two clocks is measured. The upper clock is then preset to d/c and paused, which is the time required for a light-speed signal to travel between the left two clocks. When the lower left clock strikes zero it emits a flash. When this flash is

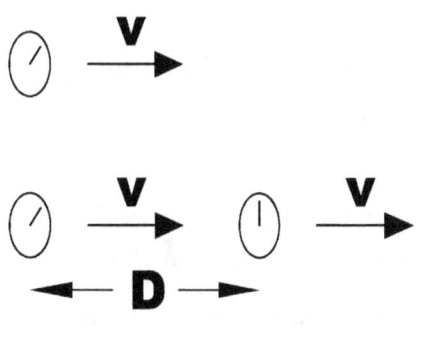

Figure A8-2

received by the upper clock it starts running from its preset value. Alternately, we could use the "flash in the middle" method to synchronize the left two clocks, then use the WWV method to synchronize the lower two clocks. Or, we could use the WWV method on both tail clocks with the signal coming from the nose clock. Further, the WWV method could be used to synchronize all three clocks by knowing each clock's distance from a flash.

For accurate time setting here on Earth, one must know the distance d, from one's location to the WWV transmitter which is located in Fort Collins, CO. An adjustment is made on the clock so that it reads d / c ahead of the time signals that the clock receives.

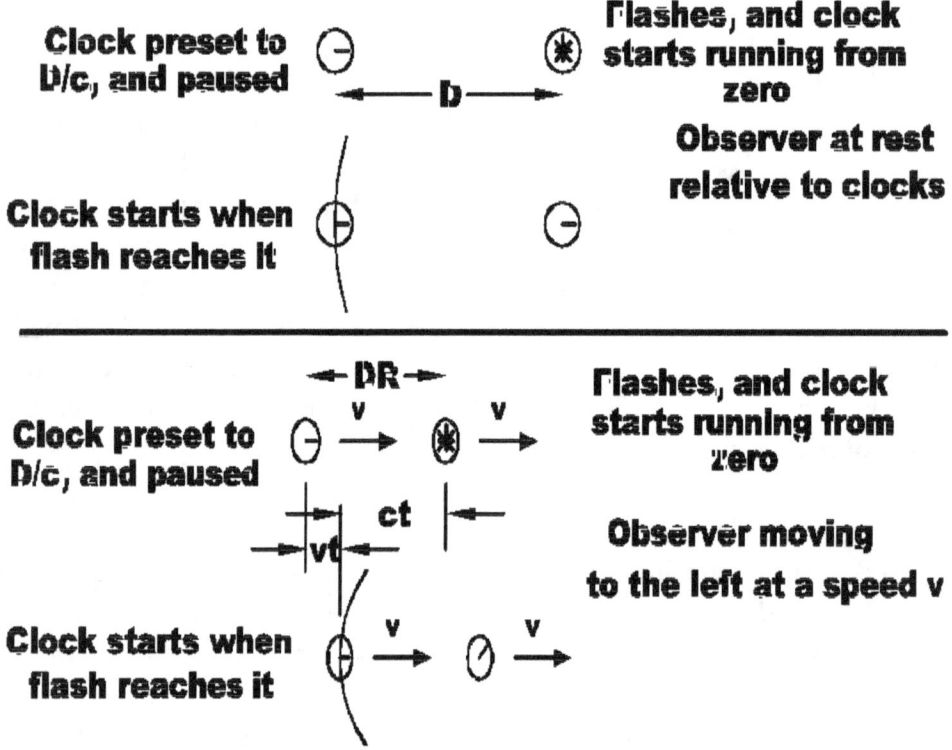

Figure A8-3 The upper half shows the WWV clock synchronization method from a frame at rest relative to the clocks. The lower half shows the same process from a frame moving to the left at a speed v relative to the clocks.

Let us further analyze the WWV clock-synchronization method and verify Equation 2-1 that, as determined by a moving observer, the following clock reads ahead by Dv/c^2 if the synchronization is done by the WWV method. To accomplish this, consider the WWV method shown in Figure A8-3. The synchronization process as determined in the rest frame of the clocks was described earlier. This same synchronization process as determined by an observer (us) who is moving to the left at a

142

speed, v, is shown in the lower portion of Figure A8-3. In our frame there are three relativistic effects to consider: 1) The separation between the two clocks, D, is contracted by the relativity factor to DR. 2) Light (and radio waves) still moves at the speed, c, in any frame, and 3) The clocks will run slow by R. Let t be the time in "our" frame for the light flash to reach the left clock (the one shown in the lower half of Figure A8-3 where the observer is moving to the left with a speed v relative to the clocks). By analysis that we have seen several times before, the light flash will move toward the left clock a distance ct, and the left clock will move toward the light pulse (to the right) a distance vt. These two distances will add to the contracted separation between the clocks, DR, see Figure A8-3. That is: $ct + vt = DR$.

Solving for t:

$$t(c + v) = DR \implies t = \frac{DR}{c + v} \qquad \text{(A8-1)}$$

where we factored t from the two terms on the left side; then divided both sides by $c + v$.

During this time t (to us) the leading clock will advance by only t times R, that is by:

$$\frac{DR^2}{c + v} = \frac{D(1 - v^2/c^2)}{c + v} = \frac{D/c^2(c^2 - v^2)}{c + v} = \frac{D/c^2(c + v)(c - v)}{c + v} = \frac{D}{c^2}(c - v)$$

where we multiplied Equation A8-1 by R; wrote out R^2; multiplied and divided ( ) by c^2; factored the binomial in ( ) (Appendix A2-2); then divided numerator and denominator by $c + v$.

$$\frac{D}{c} - \frac{D}{c^2}(c - v) = \frac{D}{c}\left(1 - \frac{c - v}{c}\right) = \frac{D}{c}\left(\frac{c - c + v}{c}\right) = \frac{Dv}{c^2}$$

At the moment that the flash reaches the left (the following) clock, it reads its preset value, D/c. The difference between the readings of the two clocks is then (following clock minus leading clock):

where we wrote the left clock reading minus right clock reading; factor D/c from each term; combine the terms in ( ) using a common denominator of c (see Appendix A1-2); perform the indicated subtraction in the numerator; multiply the two fractions together.

This reproduces Equation 2-1 that the following synchronized clock will read ahead of the leading clock by Dv/c^2, where D is the separation between the two clocks along the line of motion in the rest frame of the clocks. The fact that we obtain the same result even though we used a different synchronization method should give us confidence that we have not made errors in our development of relativity so far.

Equation 2-1 will also be reproduced by having the following clock produce the flash. An interesting exercise is to paraphrase the previous figure and analysis to show that this result still holds. [All that must be done is to replace v with $-v$ (The observer is moving in the opposite direction). This will reverse which clock (flash or not) becomes the following clock.]

The WWV method is a more practical method of synchronizing clocks than the flash-in-the-middle method, especially when more than two clocks are involved. One simply measures the rest distances from the flash device to each clock and presets each paused clock to its signal transit time from the flash. Each clock then starts when the flash or radio signal is received.

Appendix 9: Velocity Combining Simplifications in Section 2-3:

To obtain Equation 2-3 we begin with the equation from adding lengths in Figure 2-6:

$$V \Delta t + v \Delta t = DR$$

Factor Δt out of the left side; then divide each side of the equation by $V + v$:

$$\Delta t \left(V + v \right) = DR \quad \Rightarrow \quad \Delta t = \frac{DR}{V + v} \qquad (2\text{-}2)$$

To obtain Equation 2-5 we begin with Equation 2-4:

$$v' = \frac{D}{\dfrac{DV}{c^2} + \dfrac{DR^2}{V + v}} \qquad (2\text{-}4)$$

D divides out of the primary numerator and denominator (showing that the separation between the clocks does not matter as long as it is not zero). (below): Remember that in this gedanken experiment R is calculated using (upper case) V, the speed of the ship as determined by us. So, insert $R^2 = \left(1 - V^2/c^2 \right)$; we then add the two secondary fractions in the primary denominator as in Appendix A1-2:

$$v' = \frac{1}{\dfrac{V}{c^2} + \dfrac{R^2}{V + v}} = \frac{1}{\dfrac{V}{c^2} + \dfrac{1 - V^2/c^2}{V + v}} = \frac{1}{\dfrac{V \left(V + v \right) + c^2 \left(1 - V^2/c^2 \right)}{c^2 \left(V + v \right)}}$$

144

(Below) Multiply the primary numerator and primary denominator by $c^2(V+v)$; multiply out the terms in the denominator; the first and last terms in the denominator add out (cancel); divide numerator and denominator by c^2

$$v' = \frac{c^2(V+v)}{V(V+v)+c^2(1-V^2/c^2)} = \frac{c^2(V+v)}{V^2+Vv+c^2-V^2} = \frac{c^2(V+v)}{Vv+c^2} = \frac{V+v}{\frac{Vv}{c^2}+1}$$

(below): Reverse the order of the sum in the denominator:

$$v' = \frac{V+v}{1+\frac{V\!v}{c^2}} \qquad (2\text{-}5)$$

Appendix 10: Manipulations for Section 2-4: 2-D and 3-D Velocities

This is quite similar to the last manipulation in Appendix 9. Beginning with Equation 2-8 we write out R^2 in the denominator. Here R is calculated using the speed of the ship, V. (We will not write out the R in the numerator yet to keep down the clutter); then multiply the primary numerator and primary denominator by $V+v_p$:

$$v'_T = \frac{v_T\dfrac{R}{V+v_P}}{\dfrac{V}{c^2}+\dfrac{R^2}{V+v_P}} = \frac{v_T\dfrac{R}{V+v_P}}{\dfrac{V}{c^2}+\dfrac{1-V^2/c^2}{V+v_P}} = \frac{v_T R}{\dfrac{V(V+v_P)}{c^2}+1-\dfrac{V^2}{c^2}}$$

Below: We multiply out the numerator of the first secondary fraction in the denominator, then write it as two separate fractions; the first and last terms in the primary denominator subtract out; finally inserting the expression for R and rearranging the order of the terms in the denominator, we have:

$$v'_T = \frac{v_T R}{\dfrac{V^2}{c^2}+\dfrac{Vv_p}{c^2}+1-\dfrac{V^2}{c^2}} = \frac{v_T\sqrt{1-V^2/c^2}}{1+\dfrac{Vv_p}{c^2}} \qquad (2\text{-}9)$$

Appendix 11: Useful Approximations

For speeds less than about half the speed of light we can use a mathematical method known as a binomial expansion to obtain useful expressions for R, and also for $1/R$. It turns out that R is approximately $1 - \frac{1}{2}\, v^2/c^2$, and $1/R$ is approximately $1 + \frac{1}{2}\, v^2/c^2$ for low speeds. The table below shows how these approximate forms compare to the exact expressions for various speeds.

v	R	$1 - \frac{1}{2}\, v^2/c^2$	$1/R$	$1 + \frac{1}{2}\, v^2/c^2$	
0	1	1	1	1	(exact) c /100
0.99995	0.99995	1.00005	1.00005	1.00005	(1860 miles/sec)
$c/10$	0.99499	0.99500	1.00504	1.00500	(quite close)
$c/2$	0.86603	0.87500	1.15470	1.12500	(within 3%)
$0.9c$	0.43589	0.59500	2.29416	1.40500	(not close at all)

We do not need the values in this table. These values are presented to show how close these approximate expressions are to the exact expressions. What we need are the functional forms of these approximations. We can see that these functional forms do indeed give very close agreement at speeds up to nearly half the speed of light. At the speed of a jet plane, the agreement is to within one part per trillion.

Obtaining these binomial expansions is beyond the intended math level for this book. However, the table above should show that these forms are quite valid for low speeds, and approach exactness as speeds approach zero.

We will use the low-speed approximation for R to complete a proof from Section 3-5 and it is also needed in Appendix A12-2. The low-speed approximation for $1/R$ is needed in Section 4-5. In Section 3-5 we need to show is that

$$\frac{Dc}{v}\left(1 - \sqrt{1 - v^2/c^2}\right)$$

equals zero when $v = 0$. Directly plugging in zero for v yields Dc times $(1-1)/0 = 0/0$. This fraction is indeterminate – any answer will satisfy it. (Try any number, say $0/0 = 17$. To check a division one multiplies the answer (the quotient, that is, 17) times the denominator (0) and this must equal the numerator. It does, but so will any other answer, hence, it is indeterminate.

Zero speed is certainly a low speed, so let us try the approximate form for the relativity factor that is shown above, then try plugging in $v = 0$:

$$\frac{Dc}{v}\left[1 - \left(1 - \frac{1}{2}\frac{v^2}{c^2}\right)\right] = \frac{Dc}{v}\left[\frac{1}{2}\frac{v^2}{c^2}\right] = Dc\frac{v}{2c^2} = D\frac{v}{2c}$$

The 1's add out; first a *v*, then a *c* divides out of both numerator and denominator. When we insert $v = 0$ this clearly gives zero. There are calculus methods called "Limits" that can also handle this 0/0 situation.

Appendix 12: Inequalities, The Stick and Slot Paradox, and The Passing Spaceship Causality proof from Section 3-5

A12-1 Inequalities

These proofs use *inequality* manipulation. The usual rules that we have seen for equality algebra apply, but there are a few differences. The statement $a < b$ reads, "*a* is less than *b*." The point on the inequality symbol points to the lesser number. We could just as well have written $b > a$. It has the same meaning, but we would say, "*b* is greater than *a*." We will need a slight extension; $c \leq d$ reads, "*c* is less than or equal to *d*." $d \geq c$ would be read as, "*d* is greater than or equal to *c*." The rules for manipulating $\geq$ and $\leq$ are the same as for $>$ and $<$.

We can still add or subtract the same quantity to each side of an inequality without changing its validity. However, if we multiply or divide each side of an inequality by a quantity that is negative, then we must reverse the direction of the inequality. So multiplying both sides of $4 > 3$ by -1, gives $-4 < -3$. With signed numbers (positive or negative) any positive number is greater than any negative number. For example, $1 > -4$. Also, $-2 > -3$ (You have more money if your checking account is overdrawn (negative money) by \$2 than if it is overdrawn by \$3.)

Figure A12-1 shows a "Number Line." The numbers go on right and left to infinity, but to save paper (and trees) we show only the central portion. Notice that here any number to the right of another number is the larger.

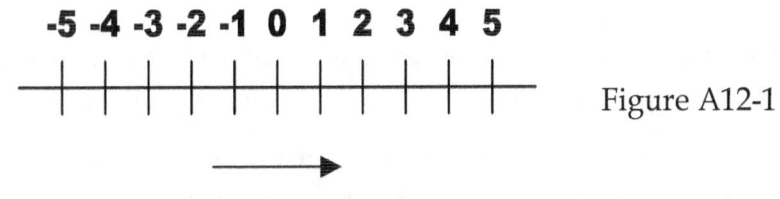

-5 -4 -3 -2 -1 0 1 2 3 4 5

Figure A12-1 A Number Line

Larger Numbers

A12-2 The Stick and Slot Paradox Solution

The paradox is described in Section 3-7. The paradox is "In the rest frame of the stick, how can the stick fall into the slot whose length is contracted?" The hint said to consider causality.

In Figure A12-2 the stick is at rest while the contracted slot is moving to the right at speed *v*. The left end of the slot hits the left end of the stick and stops. However, the right end of the slot cannot know that the left end of the slot has stopped until a signal from the left end of the slot can reach the right end of the slot. The fastest possible signals move at *c*.

As seen in the figure, the slot is contracted to a length LR. Here we will call the rest lengths of the stick and slot, L, rather than 1 meter. This will save some confusion in the simple math that follows. We will see that it makes no difference in the outcome.

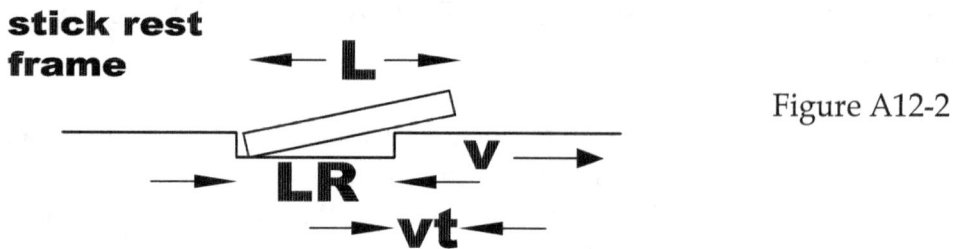

stick rest frame

Figure A12-2

Still in the rest frame of the stick: Let the time be zero when the left end of the slot stops because it hits the left end of the stick. Let t be the time when the right end of the slot reaches the right end of the stick. We will see that this will always happen, but then we already know that the stick will fall into the slot from considering the situation in the rest frame of the slot. L_{slot} and L_{stick} are both $L = 1$ meter. The subscripted L's will make it easier to see what we are doing. Until the right end of the slot receives the stop signal from the left end of the slot, it will continue to move to the right with its same speed v (causality). From the figure, the contracted length of the slot, $L_{slot} R$, plus the distance that the right end of the slot moves in the time t, vt, will equal L_{stick} (equation shown below). To solve for t, subtract $L_{slot} R$ from each side, divide both sides by v, then factor out L in the numerator. (All the L's equal 1 meter.)

$$L_{slot} R + vt = L_{stick} \quad \Rightarrow \quad vt = L_{stick} - L_{slot} R \quad \Rightarrow \quad t = L(1-R)/v$$

During this time, t, the light-speed signal from the left end of the slot will travel a distance ct. The question now is: Is ct less than or equal to L? ($ct \le L$)? If it is then the right end of the slot will pass the right end of the stick before the stop signal can reach the right end of the slot, so the stick can fall in. If they are equal, the two events will be simultaneous (a tie), so the stick will still barely fall in. The question now: Is $c L(1-R)/v \le L \Rightarrow (1-R) \le v/c$? (L divides out of the inequality, showing that it does not matter what the rest lengths of the stick and slot are, as long as they are equal to each other (and both are positive or negative together). We next divided both sides by c and multiplied both sides by v.)

The next table shows each side of this inequality for various values of v in terms of v/c. Notice that for all speeds other than $v = 0$ and $v = c$ that $1 - R$ is less than v/c. If v is either zero or c the two sides are equal. If the reader questions whether or not this inequality holds for values of v between those given in the table, or is determined to see it completely proved, the complete inequality algebra of the proof is shown below. This proof is rather tedious. No additional relativity knowledge will be developed by reading it, but math skills may be improved.

v/c	$1-R$
0	0
0.1	0.00501
0.5	0.13397
0.7	0.51000
0.9	0.56411
0.99	0.85893
0.999	0.95529
1	1

What is to be shown is that $1-R \leq v/c$ for all values of v such that $0 \leq v \leq c$ ("zero is less than or equal to v which is also less than or equal to c.")

Below: We begin with the truth that $c \geq v$. (This states the truth that the speed of light is greater than or equal to the speed of the stick or slot.); then divide both sides by c (notice that c is never negative so we do not reverse the $\geq$ symbol.); multiply each side by v/c (v is the *speed* of the stick or slot, and *does* denote the direction of movement, therefore v/c is *positive*, so again we do not reverse the $\geq$ symbol); subtract v/c from each side; then multiply through by 2.

$$c \geq v \Rightarrow 1 \geq \frac{v}{c} \Rightarrow \frac{v}{c} \geq \left(\frac{v}{c}\right)^2 \Rightarrow 0 \geq -\frac{v}{c}+\left(\frac{v}{c}\right)^2 \Rightarrow 0 \geq -2\frac{v}{c}+2\left(\frac{v}{c}\right)^2$$

To obtain the first inequality below: Subtract $\left(\frac{v}{c}\right)^2$ from each side of the last form above. Next add

$$-\left(\frac{v}{c}\right)^2 \geq -2\left(\frac{v}{c}\right)+\left(\frac{v}{c}\right)^2 \Rightarrow 1-\left(\frac{v}{c}\right)^2 \geq 1-2\left(\frac{v}{c}\right)+\left(\frac{v}{c}\right)^2 \Rightarrow 1-\left(\frac{v}{c}\right)^2 \geq \left(1-\frac{v}{c}\right)^2$$

1 to each side; the trinomial on the right side (of the second inequality below) can be written as a binomial squared: (refer to Appendix A2-1.2).

(below): We will now take the square root of both sides; but before we do that, we must verify that each side of the last inequality above is non-negative.

$$\sqrt{1-\left(\frac{v}{c}\right)^2} \geq 1-\frac{v}{c} \Rightarrow -\sqrt{1-\left(\frac{v}{c}\right)^2} \leq \frac{v}{c}-1 \Rightarrow 1-\sqrt{1-\left(\frac{v}{c}\right)^2} \leq \frac{v}{c} \Rightarrow 1-R \leq \frac{v}{c}$$

Since $v \leq c$ then $(v/c) \leq 1$ and $(v/c)^2 \leq 1$ which means that both $1-(v/c)^2 \geq 0$ and $1-(v/c) \geq 0$, so both sides are non-negative. After square-rooting, multiply each side by -1. This will reverse the inequality symbol; Finally, add 1 to each side, interchange the two terms on the right side; and recognize the relativity factor.

The final form is the inequality we need. The reader might wonder how anyone could think to do all these manipulations in this order. Actually, what is done is to start with the more complicated final expression and simplify until you arrive at something known to be true; then simply copy the steps in the reverse order. One must check to be sure that each step is still valid going backwards. Here, each backward step is valid, so we have simply presented them in the forward order.

We have yet to show that the expression is true when $v = 0$. When $v = 0$, then $R = 1$, so $1 - R = 0$ and $0 = 0 / c$. Thus, we have proved our inequality to be true. (They are equal when $v = 0$.) The stick can always fall in the slot as determined in either observer's rest frame.

A12-3 The Passing Spaceship Causality argument in Section 3-5

We must show that if $v < c$ then:

$$\frac{Dc}{v}\left(1 - \sqrt{1 - v^2/c^2}\right) < D$$

Divide both sides by D, multiply both sides by v, and divide both sides by c, and we have:

$$1 - \sqrt{1 - v^2/c^2} < \frac{v}{c}$$

which is what we had above, except that in the previous section we had $\leq$ rather than $<$. This makes no difference in the logic involved, and all the steps that we did *are* reversible, so the statement made in Section 3-5 is proven except for the case of $v = 0$. For that, see Appendix 11.

Appendix 13: Mass-Loss Manipulations and the Energy-Momentum-Mass Equation
A13-1 Mass-Loss Manipulations

Starting from Equation (4-4), we have (s is the speed of the decay particles in Frame A):

$$\frac{Ms}{\sqrt{1 - s^2/c^2}} = \frac{\dfrac{m\, 2s}{1 + s^2/c^2}}{\sqrt{1 - s'^2/c^2}} \quad where \quad s' = \frac{2s}{1 + s^2/c^2}$$

In these compound fractions (more than one fraction bar), we must be careful to keep the primary bar (the longest), and any secondary bars (less long), or even any tertiary (shorter yet) bars with their relative lengths as we manipulate. That is, in the middle expression above the primary numerator is composed of two fractions, the (secondary) shorter bar and the (tertiary) slash. The rule is, we must work with only one fraction at a time. For example: We may multiply the numerator *and*

150

denominator of any one fraction by the same thing without changing the value of the fraction. When we multiply a whole fraction by, say x, we multiply x times the numerator only. To divide a whole fraction by, say y, we *multiply* the denominator by y, or *divide* the numerator by y, (but not both).

This will be an example of rather complicated compound fractions. We will take advantage of this situation to show how compound fractions can be manipulated. To begin, divide both sides of the equation above by s, then multiply both the numerator and denominator of the primary fraction on the right side by $1 + s^2/c^2$

$$\frac{M}{\sqrt{1 - s^2/c^2}} = \frac{2m}{(1 + s^2/c^2)\sqrt{1 - s'^2/c^2}} \qquad \text{(A13-1)}$$

Let us now work only with the binomial within the square root on the right side. We must insert the expression for s' given above. We will delay taking the square root until we have simplified the expression.

$$1 - \frac{s'^2}{c^2} = 1 - \frac{\left(\frac{2s}{1 + s^2/c^2}\right)^2}{c^2} = 1 - \frac{\frac{(2s)^2}{\left(1 + s^2/c^2\right)^2}}{c^2} = 1 - \frac{\frac{4s^2}{\left(1 + s^2/c^2\right)^2}}{c^2} = 1 - \frac{4\frac{s^2}{c^2}}{\left(1 + s^2/c^2\right)^2}$$

(Above) After inserting the expression for s'; we then wrote the fraction squared as the square of numerator over the square of the denominator; squared out the $(2s)^2$; divided the numerator and denominator of the primary fraction by c^2, then multiplied the primary numerator and denominator by $(1 + s^2/c^2)^2$. (This kept our s's over our c's. In relativity, this is almost always desirable for simplifying expressions to have speeds as a fraction of c. It seems that only this ratio is meaningful.)

(below): Next we will add the two fractions by thinking of the leading 1 as the fraction 1/1. Using the method shown in Appendix A1-2, the common denominator will be the primary denominator above (times 1); square the binomial in the numerator as shown in Appendix A2-1.2; add the second and fourth terms in the numerator to obtain the third expression.

$$\frac{(1 + s^2/c^2)^2 - 4s^2/c^2}{(1 + s^2/c^2)^2} = \frac{1 + 2s^2/c^2 + s^4/c^4 - 4s^2/c^2}{(1 + s^2/c^2)^2} = \frac{1 - 2s^2/c^2 + s^4/c^4}{(1 + s^2/c^2)^2}$$

Now, as also shown in Appendix A2-1.2, we may rewrite the numerator; the expression becomes

$$\frac{(1 - s^2/c^2)^2}{(1 + s^2/c^2)^2} = \left(\frac{1 - s^2/c^2}{1 + s^2/c^2}\right)^2$$

because the square of a fraction is simply the square of the numerator over the square of the denominator. This comes from the rule for multiplying two fractions: "The product of the numerators over the product of the denominators."

We now take the square root to get back to where we started. Thus we have:

$$\sqrt{1-s'^2/c^2} = \frac{1-s^2/c^2}{1+s^2/c^2}$$

So, going back to Equation A13-1, inserting the result above; (working with the right side only) multiplying out the denominator (in the third expression); we obtain:

$$\frac{M}{\sqrt{1-s^2/c^2}} = \frac{2m}{(1+s^2/c^2)\sqrt{1-s'^2/c^2}} = \frac{2m}{\left(1+s^2/c^2\right)\left(\frac{1-s^2/c^2}{1+s^2/c^2}\right)} = \frac{2m}{1-s^2/c^2}$$

(below): Collect the first and last expressions from above; multiply each side by the denominator of the right side, then turning the equation around (right for left sides), we obtain:

$$\frac{M}{\sqrt{1-s^2/c^2}} = \frac{2m}{1-s^2/c^2} \implies 2m = M\sqrt{1-s^2/c^2} \qquad (4\text{-}5)$$

It is amazing that after all of these manipulations the result turns out to be so simple and with only our familiar relativity factor making this result different from the *classical* (pre-relativistic) law that mass is conserved, that is, $2m = M$. (pre-relativity)

A13-2 A Useful Relationship between Total Energy, Momentum, and Mass

The total energy of any object, even massless particles such as photons, is the sum of the particle's kinetic energy (KE) and its mass-energy. If we let E represent the total energy, then $E \equiv KE + mc^2$ where m is the rest mass of the object. Let us square each side of this definition:

$$E^2 = \left(KE + mc^2\right)^2 = \left(KE\right)^2 + 2\left(KE\right)mc^2 + m^2c^4 \qquad (A13\text{-}2)$$

In Section 1-10 we defined $\gamma \equiv 1/R$, that is, the reciprocal of the relativity factor. It will reduce clutter if we use this definition in our manipulations. We can then write the expression for the relativistic kinetic energy, Equation 4-8, as $KE = (\gamma - 1)\, m\, c^2$. Inserting this form into Equation A13-2 at both places yields:

$$E^2 = \left(\gamma - 1\right)^2 m^2 c^4 + 2\left(\gamma - 1\right)m^2 c^4 + m^2 c^4 \qquad (A13\text{-}3)$$

We will now work with only the first two terms on the right side of Equation A13-3. First we will factor out m^2c^4, then simplify the terms involving γ by performing the indicated square and

$$m^2c^4\left[(\gamma-1)^2+2(\gamma-1)\right]=m^2c^4\left[\gamma^2-2\gamma+1+2\gamma-2\right]=m^2c^4\left[\gamma^2-1\right] \qquad \text{(A13-4)}$$

multiplication by 2; then combine like terms.

We will now work with only the factor in the final square brackets. Using the definitions for γ and R we have:

$$\left[\gamma^2-1\right]=\frac{1}{1-\dfrac{v^2}{c^2}}-1=\frac{1-1+\dfrac{v^2}{c^2}}{1-\dfrac{v^2}{c^2}}=\frac{\dfrac{v^2}{c^2}}{1-\dfrac{v^2}{c^2}} \qquad \text{(A13-5)}$$

where we wrote out γ^2; then subtracted the two fractions by the common-denominator method described in Appendix A1-2; then the two 1's in the numerator add out.

Next, we insert the result of Equation A13-5 into A13-4, then pickup the last term in Equation A13-3; bringing m^2c^4 into the ( ), c^2 divides out of the primary numerator:

$$E^2=m^2c^4\left(\frac{\dfrac{v^2}{c^2}}{1-\dfrac{v^2}{c^2}}\right)+m^2c^4=\frac{m^2c^2v^2}{1-\dfrac{v^2}{c^2}}+m^2c^4 \qquad \text{(A13-6)}$$

Recalling the expression for relativistic momentum, Equation 4-2, we may write for the square of the momentum:

$$p^2=\frac{m^2v^2}{1-\dfrac{v^2}{c^2}}$$

Recognize this in the first term of the right side of Equation A13-6 and arrive at our final result:

$$E^2 = p^2 c^2 + m^2 c^4 \qquad \text{(A13-7)}$$

where E is the *total energy* of the object of *rest mass m*. This equation can be very useful in certain applications. Take the case of a photon (rest mass = zero): By setting $m = 0$, taking the square root of both sides, then dividing each side by c, we arrive at the relationship connecting the energy and the momentum of a photon:

$$p = \frac{E}{c} \qquad (photon) \qquad \text{(A13-8)}$$

(Do this one for yourself): Start with Equation A13-7, consider the case of a particle with a rest mass at rest (no momentum). Take the square root. Something familiar should appear.

A13-3 Showing (relativistically) that when two objects have the same momentum, the object with the lesser rest mass will have the greater kinetic energy:

(below): Begin with Equation A13-7 and subtract $m^2 c^4$ from each side:

$$E^2 - m^2 c^4 = p^2 c^2 \implies \left(E + mc^2\right)\left(E - mc^2\right) = p^2 c^2$$
$$\implies \left(KE + 2mc^2\right)\left(KE\right) = p^2 c^2 \qquad \text{(A13-9)}$$

(above): Then we factored the left side (see Appendix A2-2); then used $E = KE + mc^2$ and also the form $KE = E - mc^2$ (E is total energy).

Equation A13-9 is the equation that connects the momentum and the kinetic energy of an object with non-zero rest mass. Look at the last form of Equation A13-9 and convince yourself that for two objects that have equal momentums (equal right sides of the last form of Equation A13-9), the object with the *larger* rest mass, m, will have the *smaller* kinetic energy, and vice versa. Hint: Looking at the left side of the last form of Equation A13-9, ask yourself, "If m increased, what must be changed about the kinetic energy (KE) to keep the left side of the equation unchanged in value?" (Answer: The kinetic energy must be decreased.)

This verifies the claim made in Section 4-7, for objects moving at any allowed speed, slow or fast, that if two objects have the same magnitude of momentum, the object with the lesser rest mass will have the greater kinetic energy. Immediately after firing, the bullet will have more kinetic energy than the recoiling gun.

Appendix 14: The Relativistic Doppler Effect, Apparent Length & Apparent Speed Manipulations

A14-1 The Relativistic Doppler Effect

Before we can begin the derivation of the relativistic Doppler effect we must define a few terms. The "frequency", f, of a light wave is the number of cycles that pass by a point in space per second. Frequency is expressed in cycles-per-second, which is also called a "Hertz." The "period" of a wave, $T = 1 / f$, is the reciprocal of the frequency; it is the seconds-per-cycle of the wave. For light, the wave speed in vacuum is c. The wavelength, λ (lambda), is the distance from one wave crest to the adjacent one. Realize that each of these, except for the wave speed, c, might depend on the frame of reference.

These three quantities are related by the distance = rate times time law. The rate will be c. Notice that light will travel a distance of one wavelength, λ, in the time for one cycle, the period T. Thus we have: $\lambda = c\,T$. We can write this another way by writing it in terms of the frequency, f. We obtain: $\lambda = c / f$. Another useful expression is obtained from the previous expression by multiplying each side by f, and dividing each side by λ. We obtain $f = c / \lambda$ since $T = 1 / f$.

Consider Figure A14-1. This description will be in the rest frame of the observer. We will place zero subscripts on the variables that relate to emitter A which is at rest ($v_0 = 0$) relative to the observer. Variables related to emitter B will have no subscripts. There are two effects acting on emitter B: 1) In the observer's frame, the frequency of emitter B will be reduced by the Relativity factor, R. So, f_0 must be replaced with $f_0 R$ (This effect would be the same even if the source were

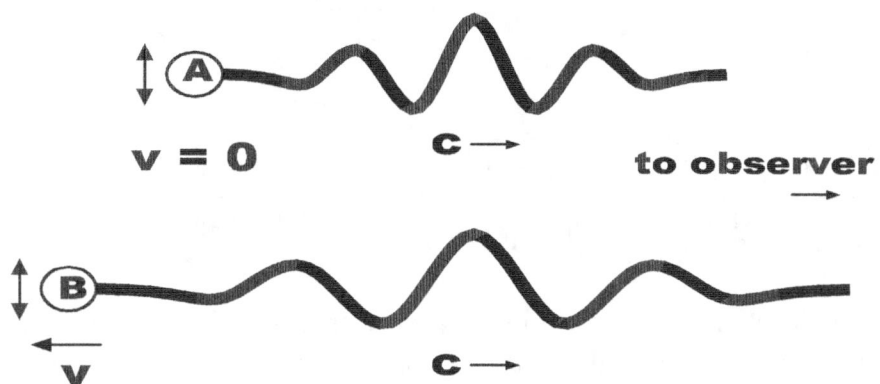

Figure A14-1 Emitter A is at rest relative to the observer. Emitter B is moving to the left at a velocity v, away from the observer.

approaching.), and 2) (The non-relativistic Doppler effect): During the time, T_0, from when emitter B emits one wave crest until it emits the next crest, the emitter will have moved away from the observer a distance of $v_0 T_0$. This will have the effect of stretching the wavelength of the emitted light from λ_0 to $\lambda = \lambda_0 + v_0 T_0$ as determined in the observer's frame. Expressing this equation in terms of the wave frequency ($\lambda = c / f$ and $T = 1/f$), we have the first equation below. Then add the two fractions on the right; reciprocate (turn over) each side; divide the denominators on both sides by c. This gives the *non-relativistic* Doppler formula.

To make the Doppler formula relativistic, all that needs to be done is to apply effect 1) from above, that is, time dilation. f_0 must be replaced by $f_0 R$.

$$\frac{c}{f} = \frac{c}{f_0} + \frac{v_0}{f_0} \implies \frac{c}{f} = \frac{c + v_0}{f_0} \implies \frac{f}{c} = \frac{f_0}{c + v_0} \implies f = \frac{f_0}{1 + v_0/c}$$

For the steps below: write out R; remember Corollary 1 – each frame will see the other going at the same speed, so replace v_0 with v which is now the speed of separation between the frames; multiply numerator and denominator by c (To multiply the square root by c, we must carry it into the radical as c^2.);

(second line) Factor the binomial inside the radical (see Appendix A2-2) and write the denominator as the square root of a square (see Appendix A3-1); the quotient of square roots is the square root of the quotient (see Appendix A3-2); divide numerator and denominator by $(c + v)$. This is Equation 5-1.

$$f = \frac{f_0 R}{1 + v_0/c} = \frac{f_0 \sqrt{1 - v_0^2/c^2}}{1 + v_0/c} = \frac{f_0 \sqrt{1 - v^2/c^2}}{1 + v/c} = \frac{f_0 \sqrt{c^2 - v^2}}{c + v}$$

$$= \frac{f_0 \sqrt{(c + v)(c - v)}}{\sqrt{(c + v)^2}} = f_0 \sqrt{\frac{(c + v)(c - v)}{(c + v)^2}} = f_0 \sqrt{\frac{c - v}{c + v}} \qquad (5\text{-}1)$$

To convert Equation 5-1 to the form for wavelengths, rather than frequencies: write the frequencies in terms of the wavelengths, λ; divide both sides by c, and reciprocate (invert both sides of the equation). This is Equation 5-2.

$$f = f_0 \sqrt{\frac{c - v}{c + v}} \implies \frac{c}{\lambda} = \frac{c}{\lambda_0} \sqrt{\frac{c - v}{c + v}} \implies \lambda = \lambda_0 \sqrt{\frac{c + v}{c - v}} \qquad (5\text{-}2)$$

If the source is *not* moving straight toward or straight away from the observer, then in the third form of Equation. 5-1 above, v in the relativity factor will be the total speed of the source, and v in the denominator will be the component (see Appendix A4-4) of the source velocity away from (+), or toward (−) the observer.

A14-2 Apparent Length Manipulation:

Equation 6-2 has the first fraction below: We move c inside the square root as c^2, and multiply it times both terms in the binomial; factor the numerator (Appendix A2-2), and include the denominator in the square root; divide numerator and denominator by $c + v$.

$$\frac{c\sqrt{1 - v^2/c^2}}{c + v} = \frac{\sqrt{c^2 - v^2}}{c + v} = \sqrt{\frac{(c + v)(c - v)}{(c + v)^2}} = \sqrt{\frac{c - v}{c + v}}$$

A14-3 Apparent Speeds Manipulation:

Begin with Equation 6-4:

$$v_{apparent} = \frac{L\sqrt{\dfrac{c - v}{c + v}}}{LR/v} = \frac{v\sqrt{\dfrac{c - v}{c + v}}}{\sqrt{1 - v^2/c^2}} = \frac{v\sqrt{\dfrac{c - v}{c + v}}}{\sqrt{\dfrac{c^2 - v^2}{c^2}}} = \frac{v\sqrt{\dfrac{c - v}{c + v}}}{\dfrac{\sqrt{c^2 - v^2}}{c}}$$

$$= \frac{cv\sqrt{\dfrac{c - v}{c + v}}}{\sqrt{(c + v)(c - v)}} = cv\sqrt{\frac{\dfrac{c - v}{c + v}}{(c + v)(c - v)}} = cv\sqrt{\frac{1}{(c + v)^2}} = \frac{cv}{c + v}$$

(first line above): Divide numerator and denominator by L, multiply numerator and denominator by v, write out the relativity factor R; use c^2 as the common denominator to add the two terms in the square root in the primary denominator; factor the c^2 out of the lower square root as c.

(second line above); multiply primary numerator and denominator by c, and factor the binomial in the square root in the denominator; combine the primary numerator and primary denominator into one square root; divide numerator and denominator by $c - v$; take the square root of the squared ( ) in the denominator (see Appendix A3-1).

Appendix 15: The Flat Space-Time Metric

Equation 5-4 was described as the four-dimensional flat space-time, or zero gravity metric. It was also said to be like the Pythagorean Theorem in four dimensions. We also said that for time to be on an equal footing with the three space coordinates, it must be multiplied by c. We wrote:

$$ds^2 = dx^2 + dy^2 + dz^2 - c^2 dt^2 \qquad (5\text{-}4)$$

where ds is an infinitesimal distance in space-time, dt is an infinitesimal interval of time, and the first three terms are infinitesimal distances in the three space directions. "OK," you say, "But why is there a minus sign in the Pythagorean Theorem?" This has been done to make ds^2 an "invariant." This means that this quantity will remain unchanged in value when one considers the same situation in reference frames that are moving relative to each other at constant speeds. (see Appendix A17-2) This is special relativity: No gravity, no accelerating reference frames. Objects within a frame can still accelerate, but not observers. The right side of Equation 5-4 can be recognized as the causality form: $D^2 - c^2 t^2$.

To illustrate the amazing power contained in Equation 5-4, let us apply it to the case of some type of clock, not necessarily a light clock, that is at rest in the "primed frame," but is moving in some constant direction at the constant speed v in the un-primed frame. Let us orient our reference frames so that all the motion is parallel to the x and the x' axes. Since there is no motion in the directions perpendicular to the x-axes, then the dy and dz terms will be zero in both frames. That is, this is a 1-D experiment.

Since ds^2 does not change (is invariant) in changing from one reference frame to another, we may write:

$$dx'^2 - c^2 dt'^2 = dx^2 - c^2 dt^2$$

Since the clock is at rest in the primed frame, then there will be no change in x', the position of the clock. Therefore, $dx' = 0$. We then have:

$$-c^2 dt'^2 = dx^2 - c^2 dt^2$$

Divide both sides by dt^2 we then have:

$$-c^2 \left(\frac{dt'}{dt} \right)^2 = -c^2 + \left(\frac{dx}{dt} \right)^2$$

The fraction dx/dt is a change in the position of the primed frame divided by the time interval in the un-primed frame. Thus, this is the speed, v, that the primed frame, and the clock, are moving in the un-primed frame. Thus, recognizing v; dividing each side by $-c^2$, and rearranging the order of the terms of the right side, gives us the first equation below:

$$\frac{dt'^2}{dt^2} = 1 - \frac{v^2}{c^2} \Rightarrow dt'^2 = dt^2\left(1 - v^2/c^2\right) \Rightarrow dt' = dt\sqrt{1 - v^2/c^2}$$

$$t' = t\sqrt{1 - v^2/c^2} \qquad (1\text{-}2)$$

(above) Multiply both sides by dt^2, take the square root of each side, and recognize the relativity factor. When we analyzed the transverse light clock in Section 1-9, the t's were the time intervals between ticks. Here the dt time intervals can be as long as the velocity between the two frames remains constant, and the clock remains at rest in the primed frame, which *is* the situation, so we replace them with t's :

This is the time dilation equation that we derived by considering the Transverse Light Clock in Section 1-9. From here we went on to length contraction and everything else that we have developed. It should not be surprising that Equation 1-2 can be derived from the flat space-time metric since Equation 1-2 was used in the derivation of the form of the right side of Equation 5-4 in Appendix A17-2.

We did not develop relativity is this mathematical way, but rather by analyzing well chosen gedanken experiments. It is much easier to gain an intuitive grasp of relativity the way that we have developed the subject.

Sometimes we claim that to put time on an equal footing with space, that t must be multiplied by ic rather than just c, where i is the square root of -1. (See Appendix A3-2) A small reason that this is done is so that the needed minus sign in the metric comes about more "naturally." The primary reason is to give the invariance characteristic to Equation 5-4. Again, "invariance" means that this expression, Equation 5-4, will compute to the same value (number of meters squared) in any reference frame. This can be very useful in some situations.

Appendix 16: Verification of the Relativistic Momentum Formula

The weakness in our discovery of the expression for momentum in section 4-2 when relativity was used was that we did not consider the case where the astronauts threw the balls out their windows at a speed that could be nearly c. Instead of re-deriving Equation 4-2 assuming any speeds, we will take Equation 4-2 and show that it does, in fact, give the expression for momentum that is conserved for any allowed speeds for v and/or V.

Examine Figure A16-1. We will transform the motion of ball *b* from Frame B to Frame A. We will then insert the velocity components of ball *b* into Equation 4-2 and show that this gives an equal, but opposite momentum to that of the momentum of ball *a* in Frame A. This will prove that Equation 4-2 *is* the expression for momentum that is conserved under relativity.

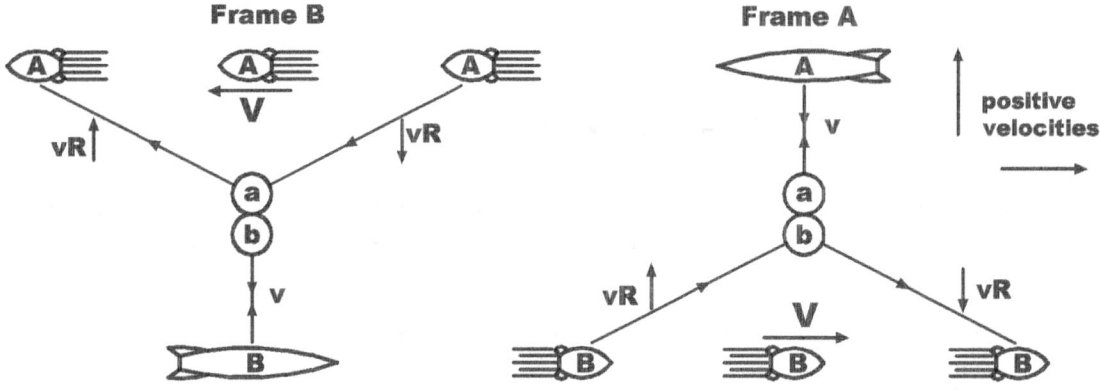

Figure A16-1 The balls thrown by astronauts from the two passing spaceships as determined in the rest frame of each spaceship

In Frame B: (We will define velocities to be positive to the right and also upward.)
$v_p = 0$ (ball *b* velocity *parallel* to the line of motion of the two frames (the two ships))
$v_T = v$ (ball *b* velocity *transverse* to the line of motion of the two frames, as in Section 2-4)
$-V$ (This is the velocity of Frame A in Frame B. This is the *variable V* in Equation 2-11.)

Recalling Equation 2-11 from Section 2-4: The primed variables on the left sides refer to Frame A. All of the un-primed variables in the right sides of the equations refer to Frame B. Inserting the values above gives the final expression for the upward (transverse) component of the velocity of ball *b* in Frame A. [Squaring $(-V)$ gives (V) squared.]

$$v'_T = \frac{v_T \sqrt{1 - V^2/c^2}}{1 - \frac{Vv_P}{c^2}} \quad \text{(2-11, repeated)} \qquad v'_T = \frac{v\sqrt{1 - (-V)^2/c^2}}{1 - \frac{(-V)(0)}{c^2}} = v\sqrt{1 - V^2/c^2}$$

In Frame A: The horizontal component of the velocity of ball b is clearly $+V$. Combining the two perpendicular velocities, v'_T and $+V$ using the Pythagorean Theorem, we obtain the equation below [See Appendix A4-1. The total speed (squared) of ball b will be the hypotenuse (squared) of the triangle]; inserting the result (squared) of v'_T from immediately above gives:

$$v_b^2 = v'^2_T + V^2 = v^2\left(1 - V^2/c^2\right) + V^2$$

for the total speed of ball b in Frame A (squared).

Recalling Equation 4-1 or Equation 4-2 for the relativistic momentum (Equation 4-1 gives any one of the components of momentum. This is all we need here since we are dealing only with the vertical components of momentum):

$$p_{upward} = \frac{m\,v_{upward}}{\sqrt{1 - v^2_{total}/c^2}} \qquad \text{(4-1, applied)}$$

Still in Frame A, let p_b be the upward component of the momentum of ball b (in Frame A). Inserting the values that we have into Equation (4-1, applied) gives the first equation below:

(second expression in first line) we have grouped the last V^2 (over c^2) in the denominator with the leading 1 in the denominator.

$$p_b = \frac{mv\sqrt{1 - V^2/c^2}}{\sqrt{1 - \dfrac{v^2\left(1 - V^2/c^2\right) + V^2}{c^2}}} = \frac{mv\sqrt{1 - V^2/c^2}}{\sqrt{\left(1 - \dfrac{V^2}{c^2}\right) - \dfrac{v^2\left(1 - V^2/c^2\right)}{c^2}}}$$

$$= \frac{mv\sqrt{1 - V^2/c^2}}{\sqrt{\left(1 - \dfrac{V^2}{c^2}\right)\left(1 - v^2/c^2\right)}} = \frac{mv\sqrt{1 - V^2/c^2}}{\sqrt{\left(1 - \dfrac{V^2}{c^2}\right)}\sqrt{1 - v^2/c^2}} = \frac{mv}{\sqrt{1 - v^2/c^2}}$$

(second line above) We then factor out the parentheses containing uppercase V within the square root in the denominator (see Appendix A2-2 for information on factoring); then write the large square root in the denominator as the product of two square roots; then divide numerator and denominator by the square root containing uppercase V.

To obtain the downward component of the momentum of ball a (still in Frame A), we realize from Figure A16-1 that the downward component of the velocity of ball a is simply v, which is also the total speed of ball a (in Frame A). Using Equation 4-1, the equation that we are to confirm, for the downward component of the momentum of ball a, we obtain:

$$p_a = \frac{mv}{\sqrt{1 - v^2/c^2}}$$

This is equal to that obtained for p_b above. Since we made no mention of any limitations on the magnitude of either v or V, then this agreement confirms Equations (4-1) and (4-2) for all speeds. These are the relativistic expressions for momentum that are conserved in all reference frames.

Appendix 17: The Lorentz Transformation and Relativistic Invariance
A17-1 The Lorentz Transformation

Most quantitative treatments of relativity will develop the Lorentz Transformation right after discovering Theorem 1 (Section 1-6). Once this has been done then mathematics (including calculus) can produce all of the results that have been developed in the present treatment plus many more in the field of physics. With this much mathematical power the thinking is then largely turned over to the mathematics; the usual result is that little intuition about relativity is developed. The conclusions seem to come from, "This result happens because these equations say so."

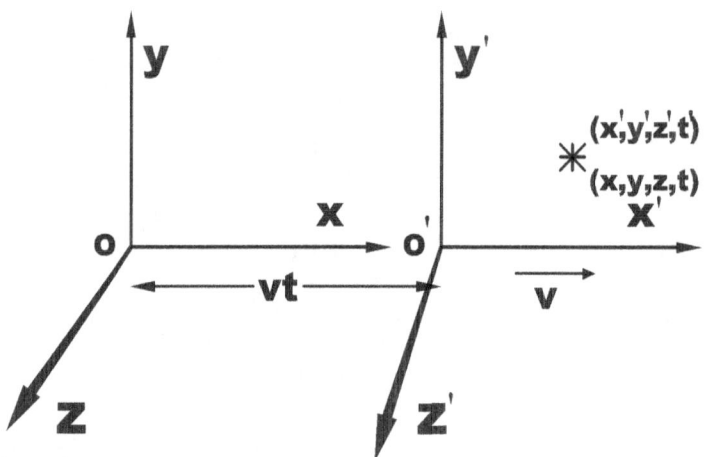

Figure A17-1 Two reference frames as shown in the rest frame of the un-primed (left) frame. The primed frame is moving along the +x-axis with a speed v. Shown at a time t measured in the un-primed frame after the two origins coincided.

We have avoided this approach because it is possible to arrive at most of the interesting results without the use of this mathematical "sledgehammer." Being able to see which relativistic effects; such as the slowing of time, the combining of velocities and clocks being out of synchronization, enter into a derivation of some new result, makes for a more rewarding experience for the reader.

However, the Lorentz Transformation will be developed here, near the end of the appendices, for two reasons. One reason is simply completeness and compatibility with other treatments to which the reader might refer. The second reason is that by using the transform we will be able to give a *mathematical* proof of an earlier claim about light spheres and causality. Unlike most treatments of relativity, the Lorentz Transformation will be derived here rather quickly by using results that we have discovered by using our gedanken-experiments approach. The results that we will use are: time dilation (the slowing of time in a moving reference frame – Equation 1-2), length contraction in the direction of motion (Section 1-13), transverse lengths (perpendicular to the motion) are not affected by motion (Section 1-12), and the asynchronization of clocks in a moving frame (Equation 2-1 and Exercise 2-2).

Figure A17-1 shows two reference frames. These frames are moving relative to each other along the *x*-axis (We may simply define the *x*-direction to be parallel to the motion, so this is not a limitation.) Some event occurs at the point (x,y,z,t) in the unprimed frame. The coordinates of this point will differ somewhat in the other frame. There the coordinates will be called (x',y',z',t'). (The apostrophe-like symbol is called "prime.") There is a clock located at the origin of each frame. At the instant that the origins coincide (move through each other), we will define the time in each frame to be equal to zero, and the clocks read zero. Before the coincidence of the origins both times were negative; afterwards, they will both be positive. We will also need two clocks that happen to be located at the position of the event, with one being at rest in each reference frame. Each of these clocks has been synchronized with the clock at the origin in its respective rest frame.

Galileo considered this situation in the 1600's (except for the differences in time, length contraction, and clock synchronization) and devised the rather obvious connections between the coordinates in the two frames of reference. These are called the "Galilean Transformation:"

$$x' = x - vt \ , \quad y' = y \ , \quad z' = z \ , \quad t' = t$$

The first relation is simply that during the time t, the origin of the primed frame will have moved to the right a distance vt, so there is a vt difference in the x-coordinates because the point of reference, the origin, has moved in one frame. These equations are equally valid as used in either frame (call either frame the primed frame), but the sign on v must be reversed when the observer changes from one frame to the other because the velocity of the other frame will then be in the opposite direction (see Figure A17-2). These equations are quite accurate until velocities become very fast.

Before learning something about relativity, these equations certainly seemed reasonable. Things get more interesting when we consider them in the light of relativity, but two of the equations do not change at all; these are the y and the z equations. These are lengths perpendicular (transverse) to the motion, and Section 1-12 demonstrated that these lengths will be the same in each frame. The other two equations will not be quite so simple and quick.

Consider the x-equation from the point of view of the *un-primed* frame as shown in Figure A17-1. In this frame all lengths parallel to the x-axis in the primed frame will be contracted by the relativity factor (Section 1-13). Let us modify Galileo's x-transform by multiplying x' (the distance between the primed origin and the event in the primed frame) by the relativity factor to contract its length; then solve for x' by dividing each side of the equation by R.

$$x'R = x - vt \implies x' = \frac{x - vt}{R} \implies x' = \frac{x - vt}{\sqrt{1 - v^2/c^2}} \qquad \text{(A17-1)}$$

So, true to its name, the relativity factor simply scales the transformation of the x's from that of Galileo. Equation A17-1 gives the value of the x'-separation of the event from the primed origin as determined by an observer at rest in the un-primed frame if she has the values of the x-position and the time of the event (x and t) as determined in the un-primed frame. If the previous sentence is confusing, then be glad that we did not develop relativity using the Lorentz transform.

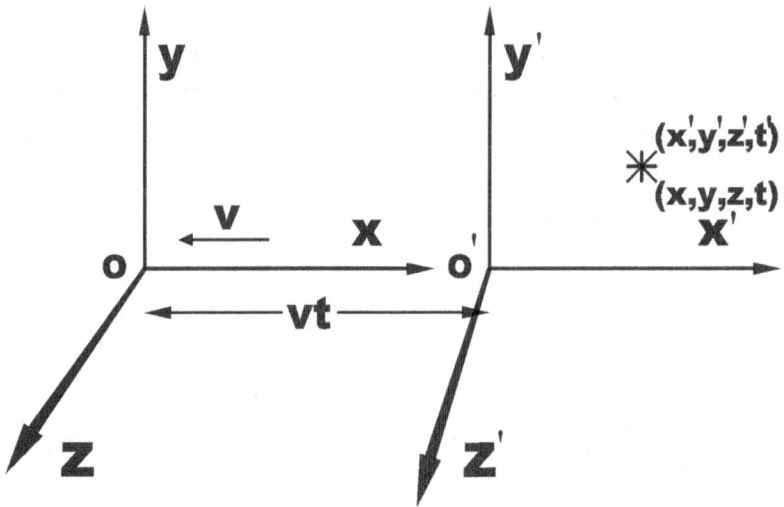

Figure A17-2 The same two frames, but shown in the rest frame of the primed system. The un-primed frame is moving to the left at a speed v.

For the time transformation, it is convenient to consider the situation in Figure A17-2, that is, from the primed frame. We know that in the primed frame clocks that are at rest in the un-primed

$$x = \frac{x' + vt'}{\sqrt{1 - v^2/c^2}} \quad , \quad y = y' \quad , \quad z = z' \quad , \quad t = \frac{t' + vx'/c^2}{\sqrt{1 - v^2/c^2}} \qquad \text{(A17-2)}$$

frame will run slow by the relativity factor, but we also know that the synchronized clocks in the un-primed frame where one is located at the origin of the un-primed frame, and the other clock is located at the event at (x,y,z,t) in the un-primed frame, will not be synchronized in the primed frame. From Equation 2-1 and Exercise 2-2, we see that the clock in the un-primed frame that is located *at the event* will be *ahead* of the clock at the un-primed origin by xv/c^2, because the event is located at the *following* clock as seen from the primed frame. Thus, we will write:

For the first equation above: The first term on the right side is the advancement of clocks (slowed by the relativity factor) in the primed frame as determined by an observer at rest in the un-primed frame, and the second term is the amount that the two clocks which were synchronized in the un-primed frame will be out of synchronization in the primed frame; we then subtract the xv/c^2 term from each side; divide each side by R, turn the equation around, and then write out R.

The complete set of four equations in the Lorentz Transformation is shown below.

$$x' = \frac{x - vt}{\sqrt{1 - v^2/c^2}} \quad , \quad y' = y \quad , \quad z' = z \quad , \quad t' = \frac{t - xv/c^2}{\sqrt{1 - v^2/c^2}} \qquad \text{(A17-3)}$$

where the primed frame is moving relative to the un-primed frame with a velocity v that is positive in the $+x$ directions. These equations tell us how to take the location and time of some event as determined in one reference frame (un-primed) and calculate the four coordinates of this same event in another reference frame (primed).

$$t = t'R + \frac{xv}{c^2} \implies t - \frac{xv}{c^2} = t'R \implies t' = \frac{t - xv/c^2}{\sqrt{1 - v^2/c^2}} \qquad \text{(A17-4)}$$

If we would like the transformation equations that will transform the coordinates in the primed frame to the equivalent coordinates in the un-primed frame, all we have to do is exchange the primed and un-primed labels, and since the un-primed frame is moving to the *left* in the primed frame, then reverse the sign on v. Since a negative number squared is equal to a positive number of the same magnitude squared, no change in the relativity factor is required. The complete inverse Lorentz transformation is given in Equation A17-4. Notice that the time transform is Equation 3-3. However, in Chapter 3 we used 1 and 2 to denote the two reference frames, while the prime usage is standard notation with the Lorentz transformation.

In summary, Equation A17-3 will transform the coordinates (position and time) of an event from their values as measured in the un-primed frame to the corresponding values of position and time as they would be measured in the primed frame if the primed frame is moving at a speed v in the $+x$ direction. Equation A17-4 does the reverse; it will use the coordinates of the same event as measured in the primed frame and compute their values as they would be measured in the un-primed frame. This provides the necessary restrictions so that we may use these two sets of equations without having to state which frame is being used for the observations, but we must use frames as defined in Figures A17-1 and A17-2.

The time transform in Equation A17-4 is the same as Equation 3-3 where we used the Bicycle race analogy. In Equation 3-3 we numbered the frames 1 & 2 rather than using primed and un-primed.

A17-2 Invariant Forms Under Lorentz Transformation:

An invariant form is an algebraic expression that will *not* change *value* when the Lorentz transformation is applied, that is, when changing from one frame to another. Let us consider the expression: $x'^2 - c^2 t'^2$. We will insert the Lorentz Transforms for x' and t' from Equation A17-3 into this form:

$$
\begin{aligned}
x'^2 - c^2 t'^2 &= \frac{(x-vt)^2}{R^2} - c^2 \frac{\left(t - vx/c^2\right)^2}{R^2} \\
&= \frac{x^2 - 2xvt + v^2 t^2}{R^2} - c^2 \frac{t^2 - 2xtv/c^2 + v^2 x^2/c^4}{R^2} \\
&= \frac{x^2 - 2vxt + v^2 t^2 - c^2 t^2 + 2vxt - v^2 x^2/c^2}{R^2} \\
&= \frac{x^2\left(1 - v^2/c^2\right) - \left(c^2 - v^2\right)t^2}{R^2} = \frac{x^2\left(1 - v^2/c^2\right) - c^2\left(1 - v^2/c^2\right)t^2}{\left(1 - v^2/c^2\right)} \\
&= x^2 - c^2 t^2
\end{aligned}
$$

Second line above: Square out the numerators of both fractions (see Appendix A2-1.2); (third line above) multiply the c^2 times the three terms in the numerator of the second fraction, collect the six terms over the common denominator, R^2; (fourth line above) the two terms beginning with "2" have added out, factor x^2 out of the first and last terms in the numerator and group these, factor the t^2 out of the third and fourth terms in the numerator and group these together; factor c^2 out of the second ( ) in the numerator; recognize the relativity factor squared, R^2 in three ( ), divide the numerator and denominator by R^2.

This shows that the form $x^2 - c^2t^2$ does not change value when the Lorentz transformation is applied, that is, when changing from one reference frame to another. Notice that we could have added $y^2 + z^2$ ($= y'^2 + z'^2$) to this expression and it would still transform without change because both y and z are unchanged in the Lorentz transformation. Thus, the form $x^2 + y^2 + z^2 - c^2t^2$ is

$$x^2 + y^2 + z^2 - c^2t^2 = x'^2 + y'^2 + z'^2 - c^2t'^2 \qquad \text{(A17-5)}$$

also invariant under the Lorentz transformation. Important point: This invariance does not mean that neither x nor t changes, only that these forms when calculated will give the same number of meters-squared for the *same event* in any reference frame. For example, both x and t might increase (or decrease), or even all four coordinates might change, but only if these forms still calculate the same number of meters-squared when transformed from one reference frame to another. Thus we may write:

For any event that occurs, this form calculates the same number of meters-squared regardless of which reference frame we use to calculate it. This is what is meant as "invariant." Notice that we do not need to write c' ("c prime") on the right side of Equation A17-7, since the speed of light is the same in all reference frames (Theorem 1).

Looking back at the manipulations that were done, it seems clear[1] that we could have substituted dx, dy, dz, and dt, where these are small changes in x, y, z, and t, and a similar thing for the primed variables, thereby arriving at the conclusion that

$$dx^2 + dy^2 + dz^2 - c^2dt^2 = dx'^2 + dy'^2 + dz'^2 - c^2dt'^2 \quad \text{(A17-6)}$$

is also invariant under the Lorentz Transformation. dx differs from Δx in that dx represents an extremely small change in x no matter what scale we are considering, while Δx refers to any size change in x. The same distinction holds for the other seven coordinates.

Equation A17-6 is the flat space-time metric equation, Equation 5-4, but here we can see the possible use of the metric equation; it is that when evaluating the form of Equation 5-4, a small change in the "location" in space-time, ds, brought about by changing position and/or time, will be the same in any reference frame if we combine dx, dy, dz, and dt (and the primed variables) in the manner of Equation 5-4. This can greatly simplify an analysis of a situation.

---

[1] This is true because all four coordinates, primed and un-primed, appear in the Lorentz transformation equations only in the numerators and also to the first power (not squared, for example). The proof of Equation A17-6 actually depends on a simple calculus method called "implicit differentiation," or some tiresome arguments equivalent to deriving the method.

In Section 3-3, using our bicycle-race analogy, we derived Equation 3-6 which demonstrates that the invariance shown in Equation A17-6 holds for even large changes in spatial and time separations between two events as long as the relative motion between the two frames remains unchanged.

A17-3 Causality Manipulations from Section 3-3

We begin with Equations (3-3), (3-4) and $D_1{}^2 = x^2 + y^2$:

$$t_2 = \frac{t_1 + xv/c^2}{R} \tag{3-3}$$

$$D_2{}^2 - c^2 t_2{}^2 = y^2 + \left(1 - v^2/c^2\right)x^2 + 2Rxvt_2 + v^2 t_2{}^2 - c^2 t_2{}^2 \tag{3-4}$$

Squaring out the middle term of Equation 3-4 and writing out R^2:

$$D_2{}^2 - c^2 t_2{}^2 = y^2 + \left(Rx + vt_2\right)^2 - c^2 t_2{}^2$$

Working with the right side only for the next five steps: Multiplying out the second term and grouping, factoring $-c^2$ from the last two terms, writing the last two terms in reverse order, and factoring $t_2{}^2$ from the last two terms:

$$D_2{}^2 - c^2 t_2{}^2 = D_1{}^2 - \frac{v^2 x^2}{c^2} + 2Rxv\left(\frac{t_1 + xv/c^2}{R}\right) - c^2 R^2\left(\frac{t_1 + xv/c^2}{R}\right)^2$$

Recognizing that the first two terms equal $D_1{}^2$, reversing the order in the numerator of the second term, and recognizing the ( ) factor in the last term as R^2, then inserting t_2 from Equation 3-3, we have:

$$D_2{}^2 - c^2 t_2{}^2 = y^2 + x^2 - \frac{v^2 x^2}{c^2} + 2Rxvt_2 - c^2\left(1 - v^2/c^2\right)t_2{}^2$$

Dividing out the R factors in the last two terms, multiplying through on the third term, then performing the square in the numerator of the last term:

Multiplying the c^2 through on the last term:

$$D_2{}^2 - c^2 t_2{}^2 = D_1{}^2 - \frac{x^2 v^2}{c^2} + 2xvt_1 + \frac{2x^2 v^2}{c^2} - c^2 t_1{}^2 - 2xvt_1 - \frac{x^2 v^2}{c^2}$$

$$D_2{}^2 - c^2 t_2{}^2 = D_1{}^2 - \frac{x^2 v^2}{c^2} + 2xvt_1 + \frac{2x^2 v^2}{c^2} - c^2 \left(t_1{}^2 + \frac{2xvt_1}{c^2} + \frac{x^2 v^2}{c^4} \right)$$

(Still working only with the right side) The second, fourth, and last terms add to zero, and the third and next-to-last terms add to zero. Thus we are left with:

$$D_2{}^2 - c^2 t_2{}^2 = D_1{}^2 - c^2 t_1{}^2 \qquad (3\text{-}6)$$

As it has happened so many times, manipulations in relativity can get messy, but seem to always simplify greatly. The relativity factor has almost magical powers in these manipulations. One needs to have the expectation that this simplification will probably occur, but if it doesn't, then go back and look for errors.

Appendix 18. Answers to Questions in the Text (that are not answered earlier in the appendices)

Section 1-9 . . . we would have detected <u>absolute</u> <u>uniform</u> <u>motion</u>.

Section 4-7 In the rest frame of the road: No. The vehicle will acquire additional mass equivalence because of its kinetic energy, but the reduction of energy in the batteries, hence reduction in the mass equivalence of the battery, will exactly compensate. The same would be true with a gasoline vehicle if we captured all emissions and internally provided the oxygen.

If we have an electric trolley which obtains it energy from external wires, then its mass and weight will increase as it speeds up, but the mass and weight of the power generating station will decrease by the same amount.

This is the essence of the statement in the next to last paragraph in Section 4-6 about energy leaving the scene. If four hydrogen atoms fused into one helium atom, and everything, including all photons and neutrinos that were produced in the reaction, remained in a box, then there would be no reduction in mass equivalence. Energy weighs! (But not much because c^2 is so large.)

Section 4-5 Showing that Equation 4-8 for the relativistic expression for kinetic energy does reduce to the classical expression, $\frac{1}{2} mv^2$, for speeds much less than c: We must use the low-speed approximation for $1/R$ from Appendix 11, that is:

$$KE = mc^2 \left\{ \frac{1}{R} - 1 \right\} \qquad \text{where} \qquad \frac{1}{R} = 1 + \frac{1}{2}\frac{v^2}{c^2} \quad (\text{low speed approx.})$$

Insert the approximation for $1/R$:

$$KE = mc^2 \left\{ 1 + \frac{1}{2}\frac{v^2}{c^2} - 1 \right\} = mc^2 \left\{ \frac{1}{2}\frac{v^2}{c^2} \right\} = \frac{mc^2}{1} \times \frac{1}{2} \times \frac{v^2}{c^2} = \frac{1}{2}mv^2$$

where we added out the 1's ; converted the first factor to a fraction and indicated the multiplications; divided a numerator and a denominator by c^2 ; we then have the desired result. (Multiplications can be done in any order – see Appendix A1-2.)

Epilogue Dictionary definition flaw: In the list of postulates, one of the last two should have been in the "can be derived from" list. See Sections 1-5 and 1-6. We used the postulate that the speed of light is not affected by the motion of its source as our Postulate 2, and then later *concluded* (using Postulates 1 & 2) that the speed of light is the same to all observers. It is possible to use either of these as Postulate 2. However, the one that we used is the easier to accept in the early stages of understanding relativity – the binary stars providing strong, and fairly easily understood supporting evidence. Using the postulate that every observer, regardless of motion, will obtain the same speed for the same light (Theorem 1) is simply too much of a first step.

Suggested Readings

Einstein, Albert. *Relativity, the Special and the General Theory*. Wings Books, 1961. This is Einstein's effort to bring relativity to the masses. A good attempt, but being first, better ways of presenting the material have since become available, so he basically states the Lorentz Transformation, then goes from there. The book is small, but probably should not be your first, or even second book on relativity.

Kaku, Michio. *Einstein's Cosmos*. W. W. Norton & Company, 2004. Primarily a history book about the life, times, and impact of Albert Einstein on the development of physics during the twentieth century. If you do not have a physics background, you may have trouble following some parts of the text.

Wolfson, Richard. *Simply Einstein: Relativity Demystified*, W.W, Norton & Company, 2003. A good descriptive-only book. Possibly more physics content than many readers will desire. Very good glossary and list of suggested readings.

Epstein, Lewis Carroll. *Relativity Visualized*. Insight Press, 1987. A very unusual relativity book. Almost all of the logic is done with rotated graphs or rolled up paper. It may be an individual issue as to whether these or equations are the more abstract. This book "derives" $E = mc^2$ by three different methods. Unfortunately, two of the methods are flawed (incorrect) because it is assumed that whatever object that he is considering to be moving has a velocity of c. Only photons can move at this speed. The third method considers only photons, so if there is value here, it is limited to only photons. The derivation in *Relativity Revealed* is longer, but it considers masses that move at any allowed speed.

Scheider, Walter. *A Serious but Not Ponderous Book on Relativity*. Cavendish Press, 2000. Quite a good book, but it assumes that the reader already knows high school algebra, a little trigonometry, and even differential calculus. He states the Lorentz Transformation, and uses it without a derivation of the transform. It has a lot of physics content, perhaps too much for the general reader. Toward the end of the book, it becomes readable only to a person with a physics background. The reader can decide whether Minkowski diagrams are less or more abstract than the equations that they portray. Very good bibliography.

Helliwell T. M. *Introduction to Special Relativity*," Allyn and Bacon, Inc, 1966. Probably the most readable of the "physics text" relativity books. Calculus-based physics is used in some places.

Bobrow, Jerry. *Cliffs Quick Review, Algebra I.* John Wiley & Sons, 2001. If the reader needs more help with the algebra than the appendices here give, the early portions of Bobrow might be of help. This inexpensive book is available in bookstores and also in large chain discount stores, especially near a college.

Feynman, Richard P. *Surely You're Joking, Mr. Feynman*, W. W. Norton & Company, 1985. This book is not about relativity, but rather about a Nobel-Prize winning physicist who crammed more living into one life than just about anybody. He did use relativity in his work, and it is refreshing to see that he was probably one of the most human of all human beings. An absolute joy to read. There is a sequel: *What Do You Care What Other People Think?* W. W. Norton & Company, 1988.

Glossary

Bolded words in the definitions refer to other entries in the glossary. "(here)" = "in this book".

Acceleration: The rate of change of **velocity** with respect to time, a **vector** quantity.

Aether: Before 1905, a hypothetical substance thought to pervade all space. The medium then thought to be the substance that wiggles when **light** passes through it.

Antimatter: A form of matter that has most its characteristics reversed, e.g., electric charge. See **positron**. **Mass** is not reversed (i.e., not negative mass) for antimatter, it is positive and equal in **magnitude** for each kind of particle and their associated anti-particles.

Big Bang: Believed by most scientists to be a gigantic explosion that began the universe (both time and space) about 13 billion years ago. This **theory** is closely tied to Einstein's **General theory of Relativity**.

Binary star: Two stars that are in orbit around each other. Very common systems.

Causality: In relativity, that if two events occur that are separated in time and space, that a light-speed signal can travel from the first event to the second event within the time separation of the two events. That is, the first event *could have caused* the second event to occur. **non-causality:** If the spatial separation is too great, and/or the time separation is too small, to allow a light-speed signal to travel between two events within their time separation. That is, the first event *could not have caused* the second event to occur.

Causality form (here): The expression, $D^2 - c^2 t^2$, where D is the separation in space between two events, and t is the separation in time between these same two events, both quantities as measured in the same **frame**. It has a negative value if causal, but has a positive value if non-causal. This form maintains the same value (is invariant) when calculated in any reference frame.

Corollary: A simple conclusion that logically follows easily or quickly from a **theorem, postulate, or hypothesis**.

Corollary 1: (here) A **corollary** that follows from **Postulate 1** that if an **observer determines** that a second observer has a certain **velocity** in a particular direction, then the second observer will determine that the first observer has a velocity of equal **magnitude**, but in the opposite direction.

Determine, Determination: Measurements taken by an **observer** if the observations have been corrected for the time differences required for light-speed signals to reach the observer from the parts of the experiment at various distances from the observer.

Doppler Effect (shift): A change in the **frequency** or **wavelength** of a **wave** caused by relative motion between the sender and receiver of the wave.

Electromagnetic radiation: Waves that consist of oscillating electric and magnetic fields that (in a vacuum) are **perpendicular** to each other and also each is perpendicular to the direction of the wave's travel.

Electron: The fundamental particle that is on the outer portion of atoms and molecules. It carries a negative charge, and its manipulation (electricity and electronics) is vital to our technology and civilization.

Energy: A lengthy discussion is required. See Section 4-4.

Frame: Short for **Reference frame**.

Frequency: The number of times per second that some event occurs. For example, the frequency of a **wave** is the number of times per second that a peak value of the wave passes by an observer.

Gedanken experiment: German for "thought" experiment. A described situation or experiment that is analyzed to arrive at some conclusion(s).

General Theory of Relativity: Published in 1916, a theory by Einstein that includes **special relativity** and explained **gravity** as being the result of **masses** following lines in **space-time** that are curved by the proximity of other **masses**.

Gravity: An attraction between all forms of matter and energy. Its strength decreases with the square of separation. Human's understanding of gravity is as yet incomplete.

Hypothesis: One or more assumptions (called **postulates**) that are assumed to be true for the sake of a logical argument to discover what logically follows from these assumptions.

Infrared (radiation) (IR): **Light** that has a somewhat lower **frequency** and consequently a somewhat longer **wavelength** than visible **light**. It is invisible to human eyes, but it can be felt as radiative warmth.

Isotropic: The same in all directions in space.

Joule: A metric unit of energy equivalent to one watt acting for one second, or a force of one Newton (about ¼ pound) acting through a distance of one meter.

Kinetic energy: The energy that a body or particle has because of its motion.

Length contraction: A fundamental relativistic effect. A length parallel to the relative motion between two observers will be determined by measurements to be shorter in any **frame** in which the length is moving.

Light: Electromagnetic waves that move through space. Usually reserved for **frequencies** that are within or nearly within the range detectible by human eyes.

Light clock: A device for measuring the passage of time that consists of either a pair of mirrors between which a pulse of **light** bounces, or a light flasher/detector and a distant mirror to return the flash to the detector which then produces an immediate flash of light. Usually used in **gedanken experiments**.

Light-year: The *distance* (not time) that a signal moving at the speed of **light** in a vacuum, c, will travel in one calendar year (365.2422 days). It is equal to about 10^{16} meters, or about 6 trillion miles or 10 trillion kilometers. This is about ¼ the distance to the nearest known star to the solar system. This is actually a system of three stars in orbit about each other, so this system (Alpha Centauri) is about 4 light-years distant.

Longitudinal (motion or separation): Motion (or separation) that is parallel to some given direction, often the motion (or separation) is parallel to the line of relative motion between two **reference frames**.

Lorentz Transformation: A group of four (or eight) equations that when given the three coordinates in space and one coordinate of time of some event will give the four coordinates of the same event in another **reference frame** which is **uniform motion** relative to the first frame. Although used extensively in most books on relativity, it is rarely used in this book.

Magnitude (of a number): The absolute value of a number, that is, remove the sign (positive or negative) of a number. For example, both +2 and -2 have a magnitude of 2.

Mass: A lengthy discussion is required. See Sec. 4-1

Massless particle: A particle such as a **photon** which when stopped or absorbed will cease to exist as a particle – only its energy and momentum survive. These never have an electric charge. Massless particles must move at c when traveling through vacuum.

Megaton: Initially, the energy released when one million tons of TNT explodes. Now rounded to 10^{15} **joules** of energy.

Michelson-Morley Experiment: An experiment preformed at several levels of sophistication on either side of the year 1900, and using several different sources of light, including sunlight and starlight, whose null (no effect) result can be interpreted as showing that light moves at the same speed regardless of the motion or orientation of the apparatus or whatever its light source. This experiment won the Nobel Prize in physics in 1907.

Neutrino: A fundamental particle that has no electric charge and which passes through matter with almost no interactions at all. Long thought to be a massless particle similar to the **photon**, but it has recently been found to possess a minute **rest mass.**

Neutron: A massive fundamental particle that has no electric charge. It is found in the nucleus of all atoms except for most hydrogen atoms. A hydrogen atom that does contain one neutron in its nucleus is called "deuterium." A radioactive form of hydrogen that contains two neutrons in its nucleus is called "tritium."

Observer: An intelligent being with sufficient apparatus to make any required measurements, logic, computations, and has the apparatus required to transmit information to others. The same, but without the intelligent being.

Paradox: A flawed argument that arrives at an illogical conclusion, a logical puzzle. For example, the sentence, "This statement is false." In relativity, a puzzle to properly apply the results of relativity to a situation that at first glance seems to produce a contradictory conclusion.

Perpendicular: At right angles to, that is, 90° apart.

Photon: The particle of **light** (electromagnetic radiation)**.** Both the energy and momentum that it carries is proportional to its **frequency**. It is its own anti-particle.

Pion: The particles in the nucleus of atoms that produces the strong nuclear force which holds the atomic nuclei together against the electrostatic repulsion of the **protons.**

Positron: The antiparticle of the **electron.** The first type of **antimatter** to be discovered.

Postulate: A statement considered to be true for the sake of logical argument. One portion of a **hypothesis.**

Postulate 1: (here) No experiment can detect absolute **uniform motion**. Only motion relative to something else has meaning.

Postulate 2: (here) The speed of **light** is not affected by the motion of its source.

Postulate 3: (here) No experiment can distinguish between uniform **gravity** and uniform **acceleration**.

Principle of Equivalence: Same as **Postulate 3**

Principle of Relativity: Same as **Postulate 1**

Proper length, Proper time: The length of an object, or separation of two events, in the **rest frame** of the entities. The time between two events in the rest frame of the events. These terms are not used in this book in an attempt to avoid the use of confusing technical jargon. These terms come from an obscure definition of the word "proper" which refers to "own self." This is definition 10 of 18 in the author's favorite dictionary which is referenced in the Epilogue. These terms are included here because many other books on relativity use them. Often, readers *incorrectly* think that these terms refer to the length or time that we are supposed to use in some calculation. See also **rest length** and **rest mass**.

Proton: A massive, positively-charged fundamental particle found in the nucleus of any atom. The number of protons in the nucleus determines which element the atom is; 1 proton produces hydrogen, 92 protons produces uranium. The other naturally-occurring elements have a number of protons that are in between these extremes.

Reference frame: A coordinate system where the location and time of some event are referred. This will be three space coordinates and one time coordinate.

Relativity factor: (here) The expression $R \equiv \sqrt{1 - v^2/c^2}$ where v is the relative speed between two **reference frames**, or the speed of some object relative to a reference frame. This factor will have a value between one (if $v = 0$) and zero for any possible speed. The author has not found this term used elsewhere.

Rest frame: The **reference frame** in which some specified object is at rest.

Rest length: The length of an object as measured in the **reference frame** in which the object is not moving. This will give the longest length compared to any other reference frame.

Rest mass: The **mass** of an object as measured in a **reference frame** in which the balance and the object are both at rest. This will give the smallest measurement of the mass compared to any other reference frame.

See: Measurements or observations taken by an observer that have *not* been corrected for the time required for a light-speed signal to reach the observer. See **Determine**.

Space-time: A four-dimensional concept combining the three space dimensions and the time.

Spatial: Refers to "space." For example, a "spatial" separation refers to the distance in space between two events, while the "temporal" separation refers to the time difference between the two events.

Special Relativity: The relativity that can be developed using only **Postulate 1** and **Postulate 2**. The work that Einstein published in 1905 which cannot deal with the observations made by an **accelerating observer**.

Supernova: The explosion produced by stars that are larger than our sun as they near the end of their lives (have exhausted their accessible hydrogen fuel) or grow too large from attracting external matter. These explosions produce the elements heavier than hydrogen and helium (such as carbon and iron) and spread them around so that later generations of star systems and planets contain them.

Theorem: A logical conclusion derived from (or proved by) a lengthy or ingenious argument.

Theorem 1: (here) Derived from **Postulate 1** and **Postulate 2,** Theorem 1 states that light moves at c relative to all **observers**.

Theory: In the scientific sense, **a hypothesis** that has been found to agree with extensive experimental evidence.

Time dilation: The relativistic result that time passes more slowly (gets stretched out) in a **reference frame** that is moving relative to the observer.

Transverse: See **Perpendicular**

Ultraviolet light (UV): Electromagnetic radiation whose **photons** have somewhat higher **frequencies** and shorter **wavelength**s than **visible light**. Ultraviolet photons often have energies great enough to break apart molecules, including organic molecules.

Uniform motion: Motion that is constant in speed and in direction.

Vector: A quantity that has both a **magnitude** (strength) and a direction in space.

Velocity: The rate of change of position with respect to time. A **vector** quantity.

Wave: A disturbance (e.g., a displacement or change in an electric or magnetic field) that travels through space.

Wavelength: The spacing between one formation in a **wave** and the position of the adjacent identical formation. For example, the distance between adjacent crests or peaks of a wave.

Subject Index

About the Author

Ray Christensen Jones 1941 – 2007, was Professor of Physics at Southwestern Oklahoma State University in Weatherford, Oklahoma for thirty-two years, 1969 – 2001. A dedicated teacher of astronomy and physics, he wrote ***Relativity Revealed: A Concrete Approach You Can Understand*** because he wanted to present an explanation of relativity that interested laymen or students could understand, even if they had little background in physics or mathematics. This book is based on a series of successful public lectures he delivered on the Southwestern campus.

Dr. Jones spent most of his childhood in Shreveport, Louisiana. He attended Louisiana State University and earned his Ph.D. in experimental physics there in 1967. He was a Post Doctoral Fellow at the University of California-Riverside for the next two years before joining the faculty at Southwestern. He taught Electronics, Digital Electronics and Microprocessors, Electricity & Magnetism, Optics, Astronomy, Solid State Physics and several other courses and laboratories over the years. In later years he did some consulting work for the 3M Corporation. Ray enjoyed bicycling, golf, traveling, flying his plane and talking to his friends. He built his own home over a period of years and enjoyed giving friends a tour, during which he could point out its superior features.

As a teacher he was known for his ability to present complex, technical and scientific concepts in an understandable form. This book is a good example of the clarity that he could bring to complex topics. His friends and colleagues felt strongly that it should be made available to the public. Special thanks are due to Jill Jones, Kinley Jones, Stan Robertson, Benny Hill and Tonya Shook for assisting with its publication.

Tributes to Prof. Ray C. Jones

Ray Jones was an inspiration to students and colleagues in many ways. He possessed a superior knowledge of physics and a marvelous ability to explain things in clear and simple language. Some of us remember him as an extraordinarily skilled experimental physicist as well as an insightful, witty and congenial colleague. Others of us remember Ray as a superb teacher and example to students. Ray was always close to his students, remaining close friends with many long after their graduation from college. He was a strong supporter, and often faculty sponsor of the national award winning SWOSU chapter of the **Society of Physics Students.** In view of his many fine qualities and lasting friendships, this is a good place to add some comments from others.

Dr. Al Harris, former president of ***Southwestern Oklahoma State University*** had this to say on the occasion of Ray's retirement: "I always tried to spend our money on teachers and students. Ray was one of my best buys. At a time when we had one of the best physics departments in the state, Ray was the best we had. He is one of the most outstanding teachers ever to teach at Southwestern. His ability to communicate physics is second to none. His intellectual abilities are among the best ever on campus. He is also a very fine individual, very dynamic and very honest. He has been very good for us and very good for the state of Oklahoma."

Dr. Benny Hill, Chairman of the SWOSU Dept. of Physics 1964 - 1990:

I was chairman of the physics department at Southwestern Oklahoma State University during 21 years of the tenure of Ray Jones there, and have many fond memories of him. Ray had a Ph.D. in low temperature experimental physics from Louisiana State University and had done a post-doctorate at the University of California at Riverside. He.was an excellent experimental physicist who always knew technology, both old and new technologies.

Ray Jones was an intellectual giant and very effective teacher. One of the graduates told me that Dr. Jones "had a really strange way of presenting material sometimes, but it was entertaining, and it seemed to be effective." That former student also commented that "Ray Jones' passing was a huge loss to the department." Ray Jones was a great experimental physicist, teacher, and mentor of SWOSU students. He had an uncommon talent for presenting some of the most complex things in physics in terms that our students could easily understand.

Ray loved to teach astronomy and did so for many years. Early on he proposed that the Department of Physics build an astronomy observatory. He designed the SWOSU Observatory and it was constructed for a mere $18,000, which included $10,000 for the dome. Then the department's 14-inch Celestron reflecting telescope was installed. To provide accurate tracking of the sky by the telescope, Ray designed, built, and installed a crystal-controlled clock drive.

The SWOSU Observatory is located near the Weatherford airport. Outside the dome, a fenced patio serves as a wind-sheltered location for additional telescope, binocular, and naked-eye

viewing. Students and visitors are guided through a wide-angle view of the skies that is not possible from within the dome.

The observatory dome now houses a 16-inch Cassegrain-focus telescope on an equatorial mount. The 16-inch-diameter objective mirror makes it possible to view many dim and distant objects which are invisible to the naked eye and to binoculars and smaller telescopes. With the rotating dome and a slot opening of over 90 degrees, views of the entire sky in 360 degrees are attained. Computer control allows for rapid aiming of the telescope to view any object that is above the horizon. An additional 8-inch Cassegrain telescope equipped to accept a camera mount is piggy-backed to the system. The equatorial mount with clock drive corrects for the Earth's rotation (which makes the stars seem to move from east to west) and eliminates the "smearing" of images on long-time photographic exposures.

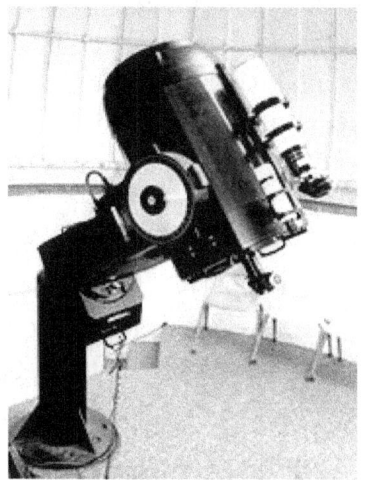

The facilities are used for regular viewing sessions scheduled in conjunction with SWOSU's Astronomy course each semester. Public viewing sessions are held by arrangement.

Another facility in which Ray Jones had a heavy hand was the machine shop in the basement of the Chemistry-Pharmacy-Physics building. As always, Ray was persistent in arguing the need for this shop. And I was persistent in pleading for this need with President Al Harris, who did not understand why a physics department needed a machine shop. One day Dr. Harris called me and said he had some unspent money from the National Science Foundation and did I have anything on which we could spend that money. I said, "Dr. Harris the only thing I can think of is a machine shop." With some anguish Al Harris agreed and Ray Jones had $2,000 left over from an NSF grant he had, which he gladly put in the pot … and we had a machine shop! Our shop was equipped with a metal lathe, a drill press, a band saw, an old Navy-surplus milling machine, and a variety of hand tools, all of which were expertly used by Ray Jones. The tribute from Carl Jantz below shows some of the utility of the machine shop.

The footprints of Ray Jones are still obvious at Southwestern Oklahoma State University. It was my privilege to see Ray make many of those footprints.

Dr. L. Dean Chapman, Professor of Anatomy & Cell Biology; Canada Research Chair, X-Ray Imaging, University of Saskatchewan, Saskatoon, Canada, (SWOSU 1975)

As you read this book, you will realize what an incredible scientist and teacher that Ray Jones was. I was Ray's student and disciple and to this day have met no one who was as creative, intuitive and caring. I can honestly say that without Ray's influence… maybe it was more like we had a lot of fun building, fixing, trying and making things work all while I learned how and why they work… I would not have gone on to graduate school and had all of the opportunities to which that led. His idea to use schlieren imaging to visualize little tornadoes made with Sears' shop vacuum cleaners was truly inspired; that directly led to an x-ray system I helped develop that is now commonly used at a number of research facilities. He was the scientist I aspire to be and after which I model the training of a new generation of scientists… it should be fun, their passion and I try to find that in them. He certainly found in me the passion for experimental physics and I thank Ray for the path he helped set for me. I should add that the training, rigor and love for physics came from an amazing pervasive spirit of the entire Physics Department at Southwestern of which Ray was a part. Certainly, Ray's legacy will live on with this beautiful, accessible book on what is considered to be a difficult and esoteric topic - relativity. For me, Ray lives on with every student I have trained, am training, and will train…but most of all within me.

Joe Abernathy, retired, Manager of the Kilby Research and Development/Silicon Fabrication Center at Texas Instruments, (SWOSU 1971).

In 1970, as a new student fresh out of the jungles of Southeast Asia, my outlook on academia was not the greatest. The first people with whom I interfaced were Dr. Ray Jones and Dr. Benny Hill, both with the physics department of Southwestern Oklahoma State College. I immediately sensed that there was something unique about these gentlemen; they genuinely cared for each and every physics student on an individual basis. I did not realize at the time that a diploma from the physics department carried with it a commitment to adopt this same philosophy for a lifetime. This philosophy required maintaining contact with the department and providing career opportunities for future graduates.

Ray required of me the same never give up attitude as that of my former Army Ranger instructors and teammates. With Ray there was no quitting due to difficulty or unforeseen problems and he instilled this in each student. This drive allowed me to achieve many goals throughout my career with Texas Instruments. I will always credit Ray with his teaching us to not flinch in the face of overwhelming adversity and to "endeavor to persevere".

Rachelle (Cole) Stephenson, 8th Grade Science, Responsive Education Solutions, Highland Village TX, (SWOSU 1992)

Dr. Ray Jones, who was affectionately known to a small group of us as "Roy", was a mentor, teacher and friend. There are so many fond memories; it's hard to share just one. Dr. Jones was able to explain complex ideas in simple words. He always had a joke to share. He would always find ways for us to experiment and get our hands dirty. I remember him taking our optics lab outside or doing some crazy experiment. He always shined during Physics Day. His classroom was the entire world. When you look at young children, you see a spark of wonder and enthusiasm for learning and experimenting that we somehow lose as we grow older. Dr. Jones was the perpetual child at heart. His endless curiosity would lead to some question that would inevitably lead him to experimenting. I will never forget how he turned my passion for tornado chasing into a personal mission to create a detection device that would help me locate them. After one of our many long conversations in his office, he giddily went home to create something that would help me with locating tornados. In a few weeks, he then invited me to his house to show me his crude, but creative lightening detection apparatus. I believe he recruited one or two of his kids to help as well. I always wondered how much of our classroom inquiries turned into family projects. His excitement turned into my excitement. He encouraged and fostered that little kid in all of us and for that I am forever grateful.

Brian Stephenson, President & COO, Tronics MEMS, Inc. Highland Village, TX (SWOSU 1991)

It is difficult for me to put into words the impact that Dr. Jones had on my education and my career. He was a terrific educator, an outstanding experimental physicist and ultimately a friend. I have several great memories of his classes (how can anyone make Electricity and Magnetism I & II tolerable?– he did) and his fantastic labs. His course syllabuses were littered with many inside jokes collected and incorporated over many years. Those who were in the department for some time grew to appreciate and look for them. He introduced me to Richard Feynman, of whom I am certain that his passion for life, physics and education, as well as his humor and mischievousness, were similar to Ray's. I have worked in high tech industry since graduating from SWOSU over 22 years ago and I have met and worked with many people with undergraduate degrees from many well respected universities such as Stanford, MIT, Cal Tech, etc. and I can honestly say that the education I received, at least at the undergraduate level, at SWOSU was on par with those schools. This is mainly because we worked directly with, and were taught directly by, many wonderfully brilliant professors. Dr. Ray Jones was right there at the top. He is and will always be a continuing inspiration for me.

Thad Gardner, Quality & Reliability Engineering Manager at Intel Corporation, (SWOSU 1990)

Who knew that when I took Digital Electronics and Electricity & Magnetism 1&2 classes from Dr. Ray Jones, it would be the start of my journey into a career into semiconductors/ microelectronics? He had a very elegant way of explaining complex concepts which enabled those of us in his classes to also understand these theories. But it wasn't only through classes that he shared his knowledge and enabled his students. Dr. Jones allowed students such as me to be involved in research projects outside of class. One of these was related to low temperature superconductors for which we wrote a proposal and received a grant to execute. I credit this as one of the key factors that allowed me to get a graduate research assistantship at Purdue University, which then led to my career at Intel. I am very privileged to have known Dr. Jones and to have been one of his many students.

Jami W. Ward Porter, BS Engineering Physics 2001, Project Engineer, Alliant Techsystems

There are a few people who helped to direct the course of my life, and Dr. Jones makes that exclusive list. I first met Dr. Jones when I was assigned to work as his student lab assistant for a basic physics course he was teaching at SWOSU. At that time I was majoring in mathematics. One day, toward the end of that semester, he simply asked, "Why aren't you a physics major?" Dr. Jones planted the ideas that made mastering physics an exciting goal and made one believe that achieving that goal was possible. I changed my major the very next day, and the course of my life was forever changed.

Dr. Jones had a way of making physics fascinating, current and applicable. Outside of class, we would ask him questions and listen to him talk for hours about various topics – he had an exemplary way of explaining concepts. His office was always full of science – things he was thinking about, working on, tinkering with or reading. In a side discussion, after class one day, I remember him opening a drawer in his office to locate an old computer mouse, so that he could explain in detail how an encoder worked. He then took apart the mouse at his own desk to compare operational differences. He was a marvelous teacher!

Some of my fondest memories of Dr. Jones are associated with his involvement with the SWOSU Physics Club. Dr. Jones had an energetic spirit! If we dreamt it, he would figure out how to make it work, developing subscale models and all. On the lighter side, he introduced us to classic movies such as, *Dr. Strangelove or: How I Learned to Stop Worrying and Love the Bomb*, and on a camping trip taught us how to hunt spiders with flashlights pressed to our heads! He was involved with every club trip and event, even opening his home to host special events – he truly cared and it showed.

Dr. Jones was brilliant with an original sense of humor. He positively influenced the lives of everyone around him and made a difference in this world because of just that. I am privileged to have known Dr. Jones, the person, and blessed to have been taught by Dr. Jones, the professor.

Craig Huffman, Atomic Layer Etch Project Manager, Sematech, Albany, NY, SWOSU 1983

I didn't appreciate the ability of Dr. Ray Jones to apply knowledge to practical matters until years after I left Weatherford. On a day to day scale the depth and breadth wasn't apparent. But over the years his abilities emerged. And it wasn't limited to oil drop experiments, the speed of light (or dark) but included everyday things, engineering projects, and personal projects; all with a humorous twist. He was a good prof who could teach and connect with the student. I recall stuff from E&M with some effort, but the E&M jokes are at the forefront. He was a humble guy with a lot of talent who would have been a great asset in the corporate world. Yet he chose to build students who in turn would influence businesses, corporations, even industries. So in the end his influence was far greater than his reach. Truly a wise man.

Logan Willis, VP eTech-WEB, Inc., BS Engineering Physics & Math. (SWOSU, 1989)

Dr. Jones was the only adviser I ever had at the college level. And he was so much more than that. He was absolutely a difference maker. His ability to deliver very complicated material in very easy to understand jargon, coupled with his wit and humor made his classes such a treasure to attend. He made learning easy! Very few reach that level of status and I was fortunate enough to be on the receiving end of his brilliance.

Kevin Johnson M.D., Program Director, Family Medicine Residency,
Gwinnett Medical Center.Lawrenceville, Georgia **(**B.S. Biophysics 1993**)**

For all of the great memories I have of Ray Jones during my time at SWOSU, it's something that happened several years later that I want to share. I literally ran into Ray in the hospital hallway in Clinton shortly after I had returned to take a position as a solo family physician in town. I was unaware he was battling cancer, and after we caught up for a few minutes on his situation, he proceed to inform me that he had heard that I had moved back, and wanted to know if I was planning to teach medical students. I told him that I was, "Just a simple country doctor" and that I was here to take care of folks, not teach.

After gently chiding me for modesty, he proceeded to tell me that I needed to be teaching. When I ask why he thought I should be a teacher, his simple retort was that I was a natural. I argued, saying he had never seen me practice medicine, and I questioned how he could possibly know if I could teach it. He informed me that he had seen me teach as a lab assistant in the physics department, and if I could distill physics concepts and teach them to my fellow undergraduates, teaching medicine had to be far easier. As we closed our short conversation in front of the Radiology department, Ray matter-of-factly pointed out that I could multiply the number of people I could help in my career if I would teach others to practice medicine rather than just practicing myself.

Ray had somehow sensed that I needed to be an educator, and that fleetingly brief conversation played a part in me accepting a position affiliated with University of North Carolina at Chapel Hill, teaching family medicine. I have since moved to Georgia, and have taken on the role of founding director for a brand new residency program in family medicine. We are actively expanding primary care training, and the physicians we graduate will each go on to provide care for thousands of patients over their careers.

I never had the opportunity to thank Ray for giving me that little nudge, and in some small way I hope these few lines can convey how much love and admiration he deserves as a mentor and as a human being.

Dr. Terry Goforth, Prof. Physics, SWOSU (SWOSU 1981)

I first met Ray when I was a high-school senior visiting college campuses. It was evident to me from the very beginning that Ray was a person who loved physics and loved sharing that knowledge with others.

Through my college years, I came to know Ray as a bright, energetic, fun-loving person. He was a talented physicist and experimentalist, and he always had one or more projects going. All who knew him know of his professional and academic skills and successes. Ray was a person who enjoyed any kind of puzzle or intellectual challenge, and he was always glad to bring colleagues and students along in his endeavor to solve a mystery. His enthusiasm for seeking new knowledge coupled with his jovial approach and quirky sense of humor made him a very popular teacher. His enthusiasm for acquiring knowledge was contagious, and class time with him was never dull. I know that many of his students, me included, count him as our favorite.

Later I came to know Ray as a colleague. He became a mentor for me. I learned much from him about teaching and motivating students and about the many skills required to be a successful member of a college faculty. I actively sought and used his advice on countless occasions. It always came down to having a clear understanding of the topic and an appropriate sense of humor. I continue to use some of his analogies, often humorous, in the classroom.

Ray's ability to see through to the heart of a problem and then simplify it was one of the reasons he excelled in teaching at all levels. This talent was apparent in his seminars on relativity which he gave at SWOSU. The talks were open to the general public, and the general public came, and came back, and kept coming back week after week. It is wonderful to have them preserved in this book.

Ray was a special, unique person who touched so many lives. Part of who I am today I owe to him, and I'm certain that this sentiment is shared by countless numbers of his students, colleagues, and friends.

Carl Jantz, Manufacturing/Product Engineer, 3M Electronic Solutions Division, Columbia, Missouri, (SWOSU 1989)

I enrolled in the Engineering Physics program in Fall of 1985. My plan was to complete the 'Pre-Engineering' program and then transfer to Oklahoma State University. Dr. Jones was assigned to be my advisor.

In the next two years, I made good friends both with classmates and with the Physics Dept staff. Dr. Jones and I conversed both as a student in his classes, as my advisor, and in general conversation. He was always approachable and would take the time to listen to a young students concerns and questions.

The spring of our sophomore year, it was time to enroll in OSU. Another Pre-Engineering student and I started the process, but after visiting the OSU campus, we realized the value of the personal interactions we had with the professors. All of the professors at SWOSU had an open door policy, and would take the time to listen to our questions, even if asked while walking down the hallway. With that, we elected to cancel the move and finish at SWOSU with the Engineering Physics program.

Over the next two years, my involvement with Dr. Jones grew as we worked on High Temperature Superconductor research and a new class created by Dr. Jones that was an 'applied' physics. The superconductor research was funded through a grant. Dr. Jones and Dr. Hill were instrumental in oversight of the writing of the proposal for the grant. Dr. Jones was the point person for the execution of the activity. He always had a way of making tasks that seemed difficult, rather simple. He was astute at using the lathe in the basement when we needed to fabricate a controlled temperature vessel we could submerge into liquid nitrogen, and then helped fabricate a temperature controller/monitoring system so that we could vary the temperature of the superconductor samples while submerged in liquid nitrogen. It wasn't pretty, but we learned a lot about applied physics and electronics, rather than shopping for a turn-key solution from a catalog.

The 'applied physics' class was a program where Dr. Jones would ask a local manufacturing company if they had any projects that they would allow us to investigate, and to fund the materials. The first project was to detect a reject mark on a web of material and sound an alarm if detected (any output that the production team could use). The biggest challenge with this project was that the reject mark was a black ink, and one of the webs was black in color. We pondered this for a bit, and found that the ink had carbon in it, which is 'visible' in the IR spectrum. So we went to the lathe/drill press and created fixturing, and with a breadboard, built our electronics circuit. The system worked great!!

The above stories are just a few examples of how Dr. Jones took the time to work with students and shared his working knowledge of physics. His teaching of 'hands on' physics has carried well into my professional career.

Mark Holmstrom, Director – Manufacturing and LSS, 3M Electronics and Energy Business Group, Austin, TX, (SWOSU 1985)

As a Physics grad, I consider myself fortunate to have had the opportunity to get know, learn and work with Dr. Ray Jones. I can honestly say that he was the most influential professor that I had at SWOSU. He had a very unique humorous style of teaching and was extraordinary to visit with one on one. His ability to help create a vision of both theoretical and applied physics in the students mind was uncanny. He truly brought the excitement to physics for me.

Tom Weichel, Jr., Wafer Fab Manager, Texas Instruments, Richardson TX (SWOSU 1990)

I am positive that I speak for many when I say that when people reflect on their time at Southwestern in the Physics Department, Ray Jones and his impact will come to everyone's list. Dr. Jones had a significant impact on me personally as my advisor, as a classroom instructor and Physics Club sponsor. I also completed an independent study course under him in Digital Electronics.

In addition to his serving as my advisor, the fact that I did an independent study under Ray created the need for me to spend a lot of one on one time with him. He was always willing to help and had an amazing depth of knowledge. Ray always encouraged me to take the full load of classes, and during the independent study, he taught me how to discipline myself for problem solving and getting things done. Time with him in and around the lab was always a learning experience.

The Physics Department at Southwestern was unique because of the access to such talented physics professors. Dr. Hill had established a staff of amazingly talented minds and experimentalists, and as students we had direct access to them which made the program unique. While I was there we had experimental activities on super conductors and we received a grant from Allied Signal to test our ideas of measuring ozone gas concentration around the Weatherford area.

Dr. Jones was always involved in either overseeing the experimental activities or authorship of the proposals. He had a brilliant mind, tons of energy, and an impact on many. His impact was significant and remains today as his students carry knowledge imparted by him along their own paths.

Dr. Garabed Armoudian, retired, Chairman, SWOSU Dept .of Physics 1990 - 2000.
Ray Jones and I were graduate students at LSU. He and I took classes together, and occasionally collaborated on class assignments. As colleagues in the Physics Department for 32 years I have watched him challenge his students and colleagues to explain physics principles in plain English words where the public can understand. He enjoyed and shared the elegance of mathematical formulation of physics principle, but to him that was not enough. He advocated that Physics principles should be presented in plain English words so that anyone can understand. In these lectures on the Special Theory of Relativity Professor Jones does exactly what he has preached for three decades.

Dr. Charles Rogers, retired, Prof. Physics SWOSU 1972 - 2012
Ray Jones was one of the best experimentalists I've known. For example, we once talked about surface currents on a wire. Ray said, "I bet I could measure that." He got a bunch of individual wires bundled together and rounded up the proper equipment and measured it.

Another time, during a demonstration experiment, he said "Suppose you shake up a bottle of pop. Everyone knows the pressure increases and it gets cold. But does it?? So he produced a bottle of pop with a pressure gauge and sealed top. He asked the audience to guess what would happen, and most of them guessed wrong. When he shook it, nothing happened.

Ray had a talent for coming up with elegant, simple ways of measuring things that took him right to the heart of the matter. That's how you win Nobel prizes, if you find an important enough subject for your experiment. I had a great respect for Ray. In a different situation, he might have earned a Nobel Prize, but he preferred to teach. He loved to see if he could make things more clear for students. He liked people so much that his office became a sort of "black hole." People who went in there tended not to come out because they enjoyed themselves and his company.

Dr. Stan Robertson, retired, Prof. Physics SWOSU 1990-2004:

Ray Jones was one of the primary reasons for my interest in a teaching job at SWOSU. I had previously become acquainted with him at professional meetings and was very favorably impressed with both his knowledge and his friendly nature. We had similar previous experiences learning digital electronics and microprocessors on our own, similar experiences as physics club sponsors and similar philosphies of teaching. I was impressed with how easily he interacted with students and with their high regard for him. Activities outside of classes, such as camping, physics club trips, or construction of floats for homecoming parades, were genuine fun for him.

Although Southwestern is properly focused primarily on undergraduate level teaching, faculty members need to remain engaged in scholarly activities. Ray used his skills as an experimentalist to continually upgrade the upper level physics laboratories. He revised the classic Millikan oil drop experiment to incorporate computer assistance. That allowed students to quickly and routinely measure the electric charge of an electron to about one percent accuracy. Prior to his work on the apparatus and procedures it had been a difficult experiment that yielded poor results. When I called it the "Millikan eye-strain experiment" he decided to fix it and succeeded beyond anything that I thought might be possible. Further, he wrote up the results and published them ("*The Millikan oil-drop experiment: Making it worthwhile.*" The American Journal of Physics 63.11 (Nov. 1995: 970).

In the last eight of my fourteen years at Southwestern, I studied astrophysics and gravity theories. In these endeavors, I was often the beneficiary of Ray's insights. He had a great intuitive grasp of these subjects, as shown here in Chapter 5 of this book. I will always be grateful for the many enlightening discussions that he provided as well as for his support when my findings sometimes ran counter to orthodox theory. That was most reassuring because I know very well that he would not have hesitated to tell me if he thought that I was wrong. Everyone should have such an able and honest friend.

www.ingramcontent.com/pod-product-compliance
Lightning Source LLC
Chambersburg PA
CBHW080242180526
45167CB00006B/2388